DeepSeek: The Complete Guide

Code, Math, and Reasoning — The Open Model That Changed the Cost Curve

Ken Tannenbaum

2026

Finnoybu Press

DeepSeek: Code, Math, and Reasoning

2026 Edition

DeepSeek — The Open Model That Changed the Cost Curve

Published by Finnoybu Press

For permissions and inquiries: press@finnoybu.org

First Edition: 2026

About the Author

Ken Tannenbaum is a technologist, entrepreneur, and AI governance researcher. He is the founder of Finnoybu Press and the founder/steward of the AEGIS Initiative, an open framework for AI safety and accountability. Learn more at aegis-initiative.com.

When he's not building governance standards for artificial intelligence, he's helping everyday people understand and use it.

Dedication

For everyone whose monthly OpenAI bill keeps creeping upward, for every CTO trying to budget the cost of AI without knowing what it'll cost next quarter, and for every engineer who suspects there has to be a cheaper way to do this.

There is. This book is about it.

The AI for Everyone Series

This book is part of the **AI for Everyone** series by Finnoybu Press, designed to help real people use AI tools effectively. Visit **press.finnoybu.org** for the latest titles and updates.

Collection One: Product Deep Dives

- *ChatGPT: The Definitive Guide (2026 Edition)*
- *ChatGPT: The Power User Guide (2026 Edition)*
- *Claude: The Definitive Guide (2026 Edition)*
- *Claude Code: The Power User Guide (2026 Edition)*
- *Gemini: The Complete Guide (2026 Edition)*
- *Gemini: The Advanced Guide (2026 Edition)*
- *Copilot: The Workplace AI Guide (2026 Edition)*
- *Copilot: The Advanced Workplace Guide (2026 Edition)*
- *Perplexity: The Research Playbook (2026 Edition)*
- *Perplexity: The Advanced Research Playbook (2026 Edition)*

Collection Two: Product Guides

- *NotebookLM: From Sources to Audio Overviews (2026 Edition)*
- *DeepSeek: Code, Math, and Reasoning (2026 Edition)*
- *Cursor: The AI Code Editor (2026 Edition)*
- *OpenClaw: Agentic AI for Everyday Work (2026 Edition)*

Collection Three: Cross-Platform Skills

- *AI Systems Playbook (2026 Edition)*
- *AI Prompt Engineering (2026 Edition)*
- *AI for Developers (2026 Edition)*
- *AI for Business (2026 Edition)*
- *AI for Students (2026 Edition)*
- *AI Governance for Practitioners (2026 Edition)*
- *Securing Agentic AI (2026 Edition)*

Table of Contents

Introduction

I wrote this book because DeepSeek is the most consequential AI release of the last two years and almost nobody outside of engineering Twitter understands what it means.

The headline coverage of DeepSeek in late 2024 and through 2025 was dominated by stock-market panic — Nvidia dropped, Microsoft's investment thesis got questioned, AI bubble talk surged. The technology underneath the headlines went largely unexplored. Most articles never got past "Chinese AI company released open-weight model that's almost as good as GPT-4 for a fraction of the cost." The real story is more interesting and more important than that summary suggests.

DeepSeek didn't just release a competitive model. DeepSeek released open-weight models that match the frontier on most workloads at roughly a tenth of the cost of the leading commercial alternatives. The economics of AI for any team paying real money to use it have shifted underneath everyone, and the implications are still working themselves out across thousands of engineering organizations and AI product roadmaps.

This book is about what DeepSeek is, how to use it, when it's the right choice, when it isn't, and what you need to think about — including the questions other coverage skips, like the China question, the geopolitical context, and the censorship realities that come with using a Chinese provider.

Who This Book Is For

This book has a wider audience than its technical-sounding subtitle implies:

- **Developers** evaluating DeepSeek for use in their applications
- **Technical leads and CTOs** building multi-model strategies and watching AI spend
- **Researchers** running large-scale prompt experiments where cost compounds fast
- **Engineers** curious about what self-hosting a frontier-class model actually requires

- **Product managers** trying to understand what's possible at what cost
- **Anyone using chat.deepseek.com** who wants to get more out of the free web app
- **Anyone curious** about the open-weights movement and what it means for AI's trajectory

You'll get the most from this book with some technical background — many chapters cover the API and self-hosting in detail. But the early chapters (web app, model lineup, the strategic story) are accessible to anyone who's used ChatGPT and wonders what the alternatives look like.

How to Read This Book

The book is organized in nine parts:

Part I: Getting Started (Chapters 1–5) introduces DeepSeek, the model lineup, the web app, the API, and your first calls in both.

Part II: The Web Chat Experience (Chapters 6–9) covers chat.deepseek.com end-to-end — context handling, thinking mode, file uploads, web search.

Part III: Building with the API (Chapters 10–15) is the developer-focused core — authentication, model selection, prompt caching (the 80% cost saver), tool calling, JSON mode, FIM completion. Working code throughout.

Part IV: Reasoning and Thinking Mode (Chapters 16–18) covers the reasoning capabilities that R1 introduced and that V4 has consolidated — when to use thinking mode, mathematical and scientific reasoning, agent workflows.

Part V: Self-Hosting and Open Weights (Chapters 19–22) covers the open-source story — Ollama for laptops, vLLM for production, BentoML for multi-GPU deployments, the distilled model lineup.

Part VI: DeepSeek for Your World (Chapters 23–26) applies DeepSeek to specific contexts — developers, researchers, cost-conscious teams, and the comparison case against GPT, Claude, and Gemini.

Part VII: Specialized Models (Chapters 27–29) covers DeepSeek-Prover-V2 (formal math), Janus (multimodal vision), and DeepSeek-OCR.

Part VIII: Safety, Ethics, and Limitations (Chapters 30–32) is the section other coverage skips — what DeepSeek gets wrong, the China question (data, censorship, trust), and the ethics of open-weight models.

Part IX: Reference (Appendices A–F) gives you an API quick sheet, 60+ prompt templates, troubleshooting, glossary, 2026 timeline, and resources.

You can read straight through or jump to whatever fits your work right now. Developers focused on the API can skim Part I, dive into Part III, and reference the rest as needed. Strategic readers can read Parts I, VI, and VIII and skim the technical middle. Self-hosters live in Part V.

A Note About Accuracy

DeepSeek changes frequently. The DeepSeek team ships new model versions, pricing changes, and capability updates on a cadence that has embarrassed even the most attentive coverage. This book is accurate as of early 2026, but specific details — exact pricing, current model versions, deprecation timelines, feature availability — may have changed by the time you read this.

The principles, techniques, and strategies have a longer shelf life than the specific numbers. How to design prompts that maximize cache hits, how to choose between thinking and non-thinking modes, how to evaluate cost-vs-quality tradeoffs — these skills carry across versions and across other providers.

When something looks different from what's described here, check DeepSeek's official documentation at api-docs.deepseek.com.

A Note About the Politics

Using a Chinese AI provider is a politically loaded choice in 2026. This book takes that seriously — we cover the China question explicitly in Chapter 31 and treat the geopolitical dimensions as a real consideration for real teams making real decisions.

The book is not pro-DeepSeek or anti-DeepSeek. It's pro-clarity. DeepSeek is a remarkable piece of engineering with real strengths and real concerns. Both deserve honest treatment.

If you're going to use DeepSeek, you should understand both. If you're not going to use DeepSeek, you should still understand both — because the open-weights movement DeepSeek represents is shaping the AI industry whether you adopt its specific models or not.

Let's Get Started

DeepSeek broke a curve that the AI industry had been treating as fixed. The frontier-class models that everyone assumed would cost $50–100 million to train cost less than $6 million. The API calls that competitors charge dollars per million tokens for cost cents. The capabilities that everyone assumed would stay locked behind major-lab APIs are now in MIT-licensed weights anyone can download and run.

That shift will keep playing out across the AI industry for years. Your job, whether you adopt DeepSeek or not, is to understand what changed.

Turn the page. Let's start there.

Part I: Getting Started

"Cheap power is the closest thing we have to a free lunch."

— Robert Solow

Chapter 1: What Is DeepSeek?

On December 26, 2024, a small AI lab in Hangzhou, China, posted an unceremonious announcement to their GitHub repository: DeepSeek-V3 was available, with weights released under the MIT license, an API priced at a small fraction of what OpenAI and Anthropic charged, and benchmark numbers that placed it among the world's strongest language models. The announcement got modest attention at first. Within a few weeks, it had triggered the largest single-day stock decline in the history of the technology sector — Nvidia shed nearly $600 billion in market capitalization on a single Monday in late January 2025 — and reset what AI was supposed to cost.

If you're reading this book, you've probably encountered DeepSeek through one of three doors. Maybe you tried chat.deepseek.com and noticed it answered almost as well as ChatGPT for free. Maybe you saw your engineering team start quietly evaluating DeepSeek for production workloads. Maybe you read one of the many "what does DeepSeek mean for the AI industry" pieces and wanted to understand what was under the hood.

Whatever brought you here, this chapter will give you a clear understanding of what DeepSeek is, how it differs from the AI tools you've been using, and what makes it consequential beyond the headlines.

What DeepSeek Is

DeepSeek is the public-facing brand of a Chinese AI research lab whose formal name is Hangzhou DeepSeek Artificial Intelligence Co. The lab was founded in 2023 by Liang Wenfeng, who also runs the quantitative hedge fund High-Flyer. The fund's GPU cluster — built up for trading algorithms over years — gave the DeepSeek team unusually direct access to compute by Chinese-AI-startup standards, and that access has shaped much of what they've built.

What DeepSeek ships, in practice, comes in three forms:

A free web chat app at chat.deepseek.com. Sign up with an email, ask questions, get answers. It looks and feels much like ChatGPT, with thinking mode, file uploads, web search, and a code interpreter. We cover the web app in detail in Chapter 3.

A commercial API at api-docs.deepseek.com. Pay per token, like every other major AI provider, but at prices roughly an order of magnitude lower than the leading Western alternatives. The API is what's powering the wave of products that have quietly migrated from OpenAI to DeepSeek over the past year. We cover the API across all of Part III.

Open-weight model releases on Hugging Face under the MIT license. Anyone can download the full model weights, fine-tune them, deploy them on their own infrastructure, or use them as starting points for further research. This has been the most consequential piece for the broader AI industry — it's hard to overstate what it means that a frontier-class model is freely downloadable. We cover self-hosting in Part V.

The same underlying models power all three. When you chat at deepseek.com, you're talking to (some version of) the same model an enterprise developer is calling via the API and a researcher is running on their own GPUs.

What DeepSeek Is Not

Understanding what DeepSeek *isn't* is important for using it well. A few common misconceptions worth clearing up early:

DeepSeek is not OpenAI in red wallpaper. It looks similar on the surface — chat interface, models with version numbers, an API with familiar parameters — but the underlying business model, training approach, distribution strategy, and political context are all distinct. Don't assume what's true about ChatGPT is true about DeepSeek.

DeepSeek is not free for everyone. The web app is free. The API is paid (though cheap). The open weights are free to download but require non-trivial compute to actually run at full size. "DeepSeek is free" is a common shorthand that hides important details.

DeepSeek is not censored across the board, but it does refuse certain topics. Like every major AI service, DeepSeek has content policies. Unlike most Western services, those policies reflect the political constraints of operating in China — questions about Tiananmen, Xinjiang, Taiwan, Hong Kong, and certain Chinese political figures are routinely declined or get scripted responses. This shows up most aggressively in the web app and in API responses going through DeepSeek's hosted infrastructure. We cover the implications in Chapter 31.

DeepSeek is not always the cheapest option for every workload. It's the cheapest *frontier-quality* option for most workloads as of early 2026, but specific use cases (very long context, certain multimodal tasks, certain languages) sometimes favor competitors. The cost calculus is workload-specific, not categorical.

DeepSeek is not a complete replacement for OpenAI or Anthropic. A multi-model strategy that uses DeepSeek for some workloads and frontier Western models for others is often the right architecture. Treating it as an across-the-board substitute usually surfaces the gaps.

A Brief History

DeepSeek's trajectory has been faster than almost any other AI lab's, and understanding the rough timeline helps explain where the product is now.

The High-Flyer Origins

DeepSeek's founder, Liang Wenfeng, ran a quantitative hedge fund called High-Flyer that had been accumulating GPUs for trading algorithms since the late 2010s. By 2022, before the launch of ChatGPT, High-Flyer had assembled what was reportedly one of the largest private GPU clusters in China — tens of thousands of Nvidia A100 cards, by some estimates. This pre-existing infrastructure put DeepSeek in an unusual position when it spun off from High-Flyer in 2023: a brand-new AI lab with first-tier compute access at a moment when most of the AI world was scrambling for chips.

V1 and V2

DeepSeek's first language models — V1 in late 2023, V2 in mid-2024 — were competent but not industry-shifting. They demonstrated capability and built up the team's expertise but didn't change the conversation. Most of the engineering decisions that would matter later — the Mixture-of-Experts architecture, attention mechanism innovations, training-cost discipline — were developed during this period without much industry attention.

V3 (December 2024)

DeepSeek-V3 changed everything. The model was a 671-billion-parameter Mixture-of-Experts (MoE) architecture that activated only 37 billion parameters per token — massive in capability and efficient in inference cost. Benchmark numbers on launch placed V3 among the strongest publicly released models in the world. The training cost — reported as approximately 2.788 million H800 GPU hours, valued at roughly $5.6 million — was a small fraction of the $50–100 million figures associated with frontier models from OpenAI and Anthropic.

Two facts about V3 caught the AI industry's attention. First, the open-weight release. Second, the cost. Together, they suggested that the assumed economics of frontier AI training had been substantially wrong — or at least that DeepSeek had figured out how to be far more efficient than the established labs.

R1 (January 2025)

A month after V3, DeepSeek released R1 — a reasoning model trained to produce explicit chain-of-thought reasoning before its final answers. R1 went head-to-head with OpenAI's o1 (the reasoning model that had been the subject of much industry buzz the previous year) on math, coding, and scientific reasoning benchmarks. By most measures it was competitive. R1 was also, like V3, released under MIT license with full weights available.

R1's release is when the broader markets noticed. The combination of "frontier-quality reasoning" plus "open weights" plus "trained for under $300K according to DeepSeek" prompted the late-January 2025 stock-market response and a wave of "AI bubble?" coverage that ran for weeks.

The V3.1, V3.2, and Specialized Releases

Through 2025, DeepSeek shipped a steady cadence of refinements: V3.1 (faster, better tool use), V3.2 (DeepSeek Sparse Attention for better long-context efficiency), V3.2-Speciale (a specialized variant), distilled models from R1 ranging from 1.5 billion to 70 billion parameters, plus the specialized lineup of DeepSeek-Prover-V2 (formal mathematics), Janus (multimodal vision), and DeepSeek-OCR (document OCR). Each release continued the pattern: open-weight, MIT-licensed, priced aggressively in the API, and aimed at maintaining or extending the cost-quality frontier.

V4 (April 2026)

DeepSeek-V4, released as a preview on April 24, 2026, is the current flagship. Two variants ship: **V4 Pro** (1.6 trillion total parameters, 49 billion active per token, 1 million token context) and **V4 Flash** (284 billion total, 13 billion active, 1 million context). V4 introduces compressed sparse attention mechanisms that handle the 1M context window efficiently. The thinking-mode capability that R1 introduced is now built into V4 as an optional per-call switch — V4 effectively supersedes R1 for most use cases. We'll cover the V4 lineup in detail in Chapter 2.

The pricing on V4 launch was characteristic: V4 Pro at roughly one-seventh the cost of Claude Opus 4.7 on coding workloads, V4 Flash at roughly one-twentieth the cost of GPT-5.5 on the same workloads. The discount through May 2026 has put effective prices even lower.

How DeepSeek Got So Cheap

A natural question, given the prices: how is DeepSeek able to charge so much less? Some combination of:

Architectural efficiency. The MoE design means most of the model's parameters are dormant on any given token. Inference cost scales with active parameters, not total parameters — V4 Pro's 1.6T total/49B active means it punches above its weight on capability while costing closer to a 49B-parameter model to run. Compressed sparse attention extends the same efficiency principle to the long-context case.

Training efficiency. DeepSeek has pushed engineering optimizations around mixed-precision training, custom CUDA kernels, communication overlap, and other systems-level work that reduces wasted compute. The publicly disclosed training cost numbers (V3 at $5.6M, R1-0528 at ~$294K) reflect these optimizations.

Compute economics. DeepSeek's GPU access through High-Flyer has historically been at lower effective cost than the cloud-rental rates Western labs typically pay. A structural advantage that's hard for competitors to replicate.

Different commercial pressure. Western AI labs (OpenAI, Anthropic) operate at scale on cloud infrastructure with substantial overhead, are funded by investors expecting commercial returns, and have to recover those costs through API pricing. DeepSeek has different funding incentives and different cost structures.

Open-weight strategy. Rather than monetizing the model itself, DeepSeek monetizes the API — but the API is a single commercial product, not the whole business model. This allows aggressive pricing in ways pure-API providers can't match.

The result: DeepSeek is structurally cheaper, not just temporarily so. The cost gap may close somewhat as Western providers respond, but it's unlikely to disappear entirely.

Where DeepSeek Fits in the AI Landscape

DeepSeek is one of several major AI options you might be using in 2026. Understanding where it fits helps you choose the right tool for each workload:

- **OpenAI (ChatGPT, GPT-5/5.5, o3, etc.)** — The market-leading commercial AI provider. Strong across most workloads, biggest ecosystem, the highest pricing. Covered in *ChatGPT: The Definitive Guide* and *ChatGPT: The Power User Guide.*
- **Anthropic (Claude, Claude Code)** — Particularly strong on long context, careful reasoning, and coding. Mid-to-high pricing, similar to OpenAI. Covered in *Claude: The Definitive Guide* and *Claude Code: The Power User Guide.*
- **Google (Gemini, NotebookLM)** — Strong in long context (1M+ tokens), deep ecosystem integration. Pricing competitive but not as aggressive as DeepSeek. Covered in *Gemini: The Complete Guide* and *NotebookLM: From Sources to Audio Overviews.*
- **Microsoft (Copilot)** — Built on OpenAI's models with Microsoft's enterprise integration. Pricing aligned with OpenAI. Covered in *Copilot: The Workplace AI Guide.*
- **DeepSeek** — The cost-quality leader. Comparable to the above on most workloads, dramatically cheaper, with the open-weights option for self-hosting.

A common pattern in production engineering organizations is **multi-model routing**: route the bulk of workloads to DeepSeek for cost efficiency, route specific workloads where Western models still have edges (very long context, certain multimodal tasks, code work where Claude Opus has particular strength) to those providers. This routing takes engineering effort to set up but compounds across the year — a 70% cost reduction on 80% of your AI bill is substantial.

We'll come back to multi-model strategies in Chapter 25 and the head-to-head comparison in Chapter 26.

What You'll Learn in This Book

This book is organized to take you from a curious user who's heard about DeepSeek to a confident user — whether you're doing serious work in the chat app, building production applications on the API, deploying models on your own GPUs, or making strategic decisions about how DeepSeek fits into your team's AI architecture:

- **Part I (Chapters 1–5)** gets you oriented: the model lineup, the web app, the API, your first calls.
- **Part II (Chapters 6–9)** covers the chat.deepseek.com experience in depth — context, thinking mode, file uploads, web search.
- **Part III (Chapters 10–15)** is the API-focused core — auth and billing, model selection in code, prompt caching (the most consequential cost lever DeepSeek offers), tool calling, JSON mode, FIM.
- **Part IV (Chapters 16–18)** dives into reasoning — when to use thinking mode, math and science workloads, agent workflows.
- **Part V (Chapters 19–22)** is for users self-hosting — the open-source story, Ollama, vLLM, BentoML, the distilled lineup.
- **Part VI (Chapters 23–26)** applies DeepSeek to specific contexts and includes the comparison case against Western providers.
- **Part VII (Chapters 27–29)** covers specialized models (Prover-V2 for math, Janus for vision, DeepSeek-OCR).
- **Part VIII (Chapters 30–32)** covers what you need to know about limitations, the China question, and the ethics of open-weight AI.
- **Part IX (Appendices A–F)** gives you reference material for ongoing use.

Every chapter is useful on its own. If you're a developer who wants the API and the cost story, skim Chapter 1, then jump to Chapter 10. If you're a strategic reader, read Parts I, VI, and VIII and skim the rest. If you want the full picture, read straight through.

Let's get started. Chapter 2 covers the model lineup — what each model is good for, when each one is the right choice, and how to think about the lineup as it continues to evolve.

Chapter 2: The DeepSeek Model Family

Most AI users encounter exactly one model from a given provider. They use ChatGPT and they're using GPT-5.something — they don't think much about it. They use Claude and they're using Sonnet or Opus and the picker is right there in the corner. The model picker is mostly invisible.

DeepSeek is different. The lineup is broader, more frequently updated, and matters more for cost and capability decisions than the lineups at OpenAI or Anthropic. A team using DeepSeek seriously will routinely switch between V4 Pro and V4 Flash for different workloads, will need to decide whether reasoning mode is worth its cost on each call, and may end up self-hosting a distilled model for specific use cases. Knowing the lineup well isn't optional — it's the foundation for getting the cost-quality tradeoffs right.

This chapter walks through every model worth knowing about as of early 2026, what each is for, and how to think about the lineup as it continues to evolve.

The Current Flagships: V4 Pro and V4 Flash

DeepSeek-V4, released as a preview on April 24, 2026, is the current flagship line. Two variants ship under the V4 banner, intended for different cost-quality positions on the same architecture:

DeepSeek-V4 Pro

The full-strength model. **1.6 trillion total parameters** in a Mixture-of-Experts architecture, with **49 billion active parameters per token**. The MoE design means the model is enormous in capability but far cheaper to run than a dense 1.6T model would be — only the relevant experts activate for each token.

Key specs: - **Context window**: 1 million tokens - **Maximum output**: 384K tokens - **Thinking mode**: built-in, optional per call - **Tool calling**: yes, including parallel tool calls - **JSON mode**: yes, with schema-style structured output - **License**: MIT - **Public release**: open weights on Hugging Face

V4 Pro's pricing as of early 2026 lands at roughly **one-seventh the cost of Claude Opus 4.7** on coding workloads and **one-sixth the cost of GPT-5.5** on the same. With the 75% discount DeepSeek has been running through May 2026, the effective gap is even larger.

Use V4 Pro for: production AI workloads where quality matters, coding agents, complex reasoning tasks, work that would otherwise go to Opus or GPT-5.5.

DeepSeek-V4 Flash

The cost-efficient sibling. **284 billion total parameters**, **13 billion active per token**. Flash is calibrated to be a step down in quality but a much larger step down in cost — typically running at roughly **one-fifth to one-tenth the cost of V4 Pro per call**.

Same context window (1M tokens), same thinking-mode capability, same tool-calling and JSON support, same MIT license.

Use V4 Flash for: high-volume workloads where some quality drop is acceptable, evaluation and experimentation, anywhere V4 Pro would be overkill.

A common architecture pattern: route the bulk of an application's traffic to V4 Flash, escalate to V4 Pro for queries that require extra capability. The cost gap is large enough that this routing moves the bottom line.

The R1 Line: Reasoning Specialists (Now Largely Superseded)

Before V4, DeepSeek's reasoning capabilities lived in a separate line of models — DeepSeek-R1, R1-Zero, R1-0528, and others. These were models specifically trained to produce chain-of-thought reasoning before their final answers, comparable to OpenAI's o1 series.

The good news for users today: V4 absorbed R1's reasoning capabilities. V4's optional thinking mode (available on both Pro and Flash variants) covers what R1 was used for. For most current workloads, you don't need to think about R1 separately — it's a per-call switch on V4.

Why R1 still matters:

- **Existing R1 deployments** still work. If your team built systems around R1, those systems aren't broken; they just have a successor available.
- **The R1 distilled models** (the smaller models trained on R1's outputs) remain the best path for self-hosting reasoning capability on resource-constrained hardware. We cover the distilled lineup in Chapter 22.
- **The R1 release was when** the AI industry took DeepSeek seriously. Knowing the history matters for understanding the lineup.

The R1 API endpoints remain available but are increasingly considered legacy. New work should use V4.

The V3 Line: The Original Frontier-Cost-Disruptor

Before V4 there was V3 — the December 2024 release that triggered the cost-curve recognition. The V3 line included:

- **V3 Base/V3 Chat** (December 2024)
- **V3-0324** (mid-2025 refinement)
- **V3.1/V3.1 Terminus** (faster, better tool use)
- **V3.2/V3.2-Exp** (introduced DeepSeek Sparse Attention for long-context efficiency)
- **V3.2-Speciale** (specialized variant)

V3 endpoints remain available and serviceable. For most current workloads V4 is the better choice, but V3 is a reasonable fallback for cost-sensitive work where the V4 capability gap doesn't justify the price difference. The legacy `deepseek-chat` API alias points at the V3.2 line for backward compatibility, with planned deprecation toward the end of 2026.

Distilled Models: 1.5B to 70B

Around the R1 release, DeepSeek published a series of **distilled models** — smaller models trained to mimic R1's reasoning behavior. These are the workhorses of self-hosted DeepSeek deployments:

- **1.5B** — Edge devices, mobile, browser inference
- **7B/8B** — Laptops with consumer GPUs
- **14B** — Higher-end laptop GPUs (24GB VRAM)

- **32B** — Workstation GPUs, single-GPU production servers
- **70B** — Multi-GPU production deployments

The distilled models are themselves open-weight under the MIT license. They give you reasoning-capable AI you can run on hardware you control — useful for privacy-sensitive workloads, latency-sensitive use cases, or situations where API dependency is a problem.

For most teams, distilled models aren't a primary deployment path — the API is cheaper and easier. But they're invaluable in specific scenarios. We cover self-hosting in detail across Part V (Chapters 19–22).

Specialized Models

Beyond the general-purpose lineup, DeepSeek ships several specialized models for specific workloads:

DeepSeek-Prover-V2

A model specialized for **formal mathematical proof generation**, integrated with the Lean theorem prover. For mathematicians and researchers working on formal verification, Prover-V2 is state-of-the-art at its niche.

Most users will never need this. For the small audience that does, it's covered in Chapter 27.

Janus

The **multimodal vision** lineage. Janus models can process images and produce both text and image outputs. As of early 2026, Janus is competitive with GPT-5.x vision and Gemini's vision capabilities for many workloads, though it's not the leader on every benchmark.

For teams already on DeepSeek who need multimodal capability, Janus is the natural fit. We cover it in Chapter 28.

DeepSeek-OCR

A specialized model for **high-accuracy document OCR**. Tuned for extracting text from documents, including layouts that general-purpose vision models tend to mangle (multi-column academic papers, complex tables, handwritten material).

For teams building document-processing pipelines, DeepSeek-OCR is often a better starting point than the general-purpose vision models. Covered in Chapter 29.

Choosing the Right Model: A Decision Tree

When you're building something new on DeepSeek, model selection is one of the first decisions. A practical decision tree:

Are you building a production application that calls the API?

- **Most of your traffic**: start with **V4 Flash** for cost efficiency
- **Quality-critical traffic**: route to **V4 Pro** for those specific calls
- **Reasoning-heavy traffic** (math, multi-step logic, agent workflows): use **V4 with thinking mode enabled**
- **Documents or images**: use **Janus** for general vision, **DeepSeek-OCR** for OCR specifically
- **Formal math or proofs**: use **DeepSeek-Prover-V2**

Are you self-hosting?

- **Laptop/consumer hardware**: pick the largest distilled model that fits in your VRAM (often 7B or 14B)
- **Workstation with one beefy GPU**: 32B distilled or, with quantization, V3.2 (with cooperation from your inference framework)
- **Multi-GPU production**: V3 or V4 Flash full-precision, served via vLLM or BentoML

Are you exploring or learning?

- Start with the **chat.deepseek.com** web app — free, hosted, no setup
- For API exploration: **V4 Flash** is the cheapest path to feel out behavior

We'll come back to model selection in code in Chapter 11, where we cover the API-side mechanics of switching between models in a single application.

The Model Lineup Will Keep Evolving

DeepSeek ships frequently. The lineup described in this chapter is accurate as of early 2026, but new models, refinements, and specialized variants appear at a faster cadence than most other AI labs ship.

Some patterns to expect:

Continued cost reductions. Each new release has tended to push prices down or hold them while improving capability. Don't assume current prices are floors.

More specialized models. Prover, Janus, and OCR are part of a broader strategy. Expect more domain-specific models — possibly for code, possibly for science, possibly for other high-value verticals.

Faster context windows. V4's compressed sparse attention pushes the cost of 1M-token context calls down. Expect the long-context game to keep evolving.

Better thinking-mode efficiency. The cost of reasoning has been dropping with each release. Watch for thinking mode becoming closer to the default for more workloads.

API consolidation. The current API has multiple model endpoints (V4 Pro, V4 Flash, R1, V3.x). DeepSeek is likely to consolidate these over time toward fewer, smarter endpoints.

When something new ships, the principles in this chapter — match capability to workload, route by cost, treat new models as candidates rather than mandatory upgrades — carry forward. The specifics change; the framework for choosing doesn't.

The next chapter turns from the model lineup to the most accessible way to use it: chat.deepseek.com, the free web app where most casual users will encounter DeepSeek for the first time.

Chapter 3: The DeepSeek Web App

Most people who try DeepSeek do so through chat.deepseek.com — the free web app that lets anyone with a browser sign up and start asking questions. The web app is the underrated half of DeepSeek's product. The API gets the engineering attention; the open weights get the AI-industry attention; the web app quietly serves a substantial daily-active user base who use it the same way they used to use ChatGPT, just with different politics and a different price.

This chapter walks through the web app end to end: signing up, the interface, the features, and the workflow patterns that get the most out of it.

Signing Up

Go to **chat.deepseek.com**. You'll see a landing page with a sign-up/sign-in flow.

You can sign up with:

- **Email and password** — works in most regions, simplest path
- **Phone number** — available in many regions, often used in mainland China
- **Google account** — available in some regions
- **WeChat** — available primarily for users in China

A few practical notes on signup:

Region matters. DeepSeek's signup flow varies by region. Users in some jurisdictions may see different authentication options or different terms of service. The product itself is mostly the same once you're signed in.

Email confirmation may be required. Standard practice — confirm via the link in your inbox before your account is fully active.

There's no free-tier paywall. Once you're signed up, the web app is free to use. There are no daily message limits to speak of (though heavy use may trigger soft rate limits during high-load periods).

The Interface

The chat.deepseek.com interface is intentionally familiar. If you've used ChatGPT, Claude, or Gemini, you'll recognize the pattern immediately:

- **Chat list** on the left (your conversation history)
- **Active conversation** in the center
- **Input box** at the bottom with a few utility buttons
- **Settings/account** in a corner

A few things that are specifically DeepSeek:

The Model Selector

A small selector in the input area lets you choose which model handles your message. Options typically include the current flagship (V4-class) and the cost-efficient variant (Flash-class). The naming on chat.deepseek.com sometimes lags the API-side naming – you may see "DeepSeek-V4" rather than "V4 Pro" depending on the build.

For most casual use, the default works fine. Switch to the more powerful variant when you have a specific reason – complex reasoning, careful writing, code that has to work.

The Thinking Mode Toggle

A toggle in the input area enables **thinking mode**, where the model produces explicit chain-of-thought reasoning before its final answer. When enabled:

- You'll see the reasoning unfold in a designated section before the final response appears
- The full response takes longer (the model is generating more tokens)
- For complex problems – math, multi-step analysis, careful coding – the answer quality is meaningfully better
- For simple questions, thinking mode is overkill and adds latency

There's a section on when to use thinking mode in Chapter 7. The short version: enable it when you'd want a person to think before answering.

The Web Search Toggle

A toggle that lets DeepSeek search the open web in real time and ground its answer in what it finds. When enabled:

- The model formulates a search query, retrieves results, and uses them in its response
- Citations appear in the response indicating which sources informed which claims
- Your answer reflects current information rather than training-data-cutoff knowledge

Web search is useful for current events, recent product launches, anything where the answer might have changed since training. We cover web search in detail in Chapter 9.

The File Upload Button

DeepSeek's web app accepts file uploads — PDFs, documents, spreadsheets, images. We cover the supported formats and their limits in Chapter 8.

What's Different from ChatGPT

For users coming from ChatGPT, a few things land differently in DeepSeek's web app:

Thinking mode is more prominent. ChatGPT hides reasoning behind specific model selections (o-series). DeepSeek puts the toggle in the input area where it's a per-message decision.

Model selection is simpler. ChatGPT's picker has many options across many tiers. DeepSeek's has two or three. For most users, fewer choices is better.

The output style is more restrained. DeepSeek produces less verbose responses than ChatGPT for the same prompt. This is configurable through prompting, but the default leans toward "answer the question" and away from "let me first contextualize."

Specific topics are unavailable. Questions about Tiananmen Square, Xinjiang, Taiwan, Hong Kong's political situation, certain Chinese political figures, and a handful of other topics produce either refusals or clearly scripted responses. This is a consistent feature of the web app and the API. We cover the implications in Chapter 31.

No persistent memory. ChatGPT's "memory" feature, which carries information across conversations, has no direct equivalent in DeepSeek's web app. Each conversation starts fresh.

No custom instructions. ChatGPT lets you set persistent custom instructions that apply to every conversation. DeepSeek's web app doesn't. For per-message persona setting, you can prefix prompts with role context, but it doesn't persist across the session.

Free vs. Paid Features

The free tier of chat.deepseek.com is free and capable. There's no premium subscription that unlocks more powerful models or higher-quality responses — both flagship and Flash models are available for free, and so is thinking mode.

What's reserved for paying customers, in the sense of "you pay if you want this":

- **The API** — the API is paid, period. Costs apply per token.
- **Higher rate limits at scale** — the web app has soft limits for very heavy users; API access bypasses them
- **Production deployment** — the web app is for individual use, not for embedding into your own product

There's no "DeepSeek Plus" subscription competing with ChatGPT Plus. The economics are different: chat.deepseek.com is a top-of-funnel product to introduce users to DeepSeek, while the API is the commercial offering.

File Uploads, Web Search, Code Interpreter

The web app supports several adjacent features that turn it from a chat interface into a working tool:

File uploads — drop PDFs, DOCX, spreadsheets, images, plain text into a conversation and ask questions about them. Multiple uploads per conversation are supported. Image uploads include OCR for documents-as-images. Covered in Chapter 8.

Web search — toggle on to ground responses in current web content with citations. Covered in Chapter 9.

Code interpreter — for some queries, DeepSeek can execute code in a sandboxed environment to produce charts, run analyses, or verify computations. The interface varies; sometimes it surfaces as "I'll run this for you" inline, sometimes as a more explicit toggle. Covered in Chapter 8.

For users coming from ChatGPT, the broad strokes of these features will be familiar. The specifics differ but the patterns are the same.

Conversation Management

The chat list on the left holds your conversation history. A few practical notes:

Conversations are persistent by default. Anything you ask is saved unless you explicitly delete it. Convenient, but worth knowing — the chat history is a record.

Renaming and organizing. You can rename conversations to make them findable later. There are no folders or tags as of early 2026, just a flat chronological list with rename support.

Search. Most builds support keyword search across your chat history. Useful when you remember roughly what you asked but not when.

Sharing. Some interface versions let you share a conversation as a public link. Treat it as you would any public link — once shared, the contents are out there.

Deletion. You can delete individual conversations. Whether this purges the data from DeepSeek's servers depends on their data retention policy; covered in Chapter 31.

Mobile Apps

DeepSeek ships native mobile apps for both major platforms:

- **Android** — available on Google Play in supported regions
- **iOS** — available on the App Store in supported regions

The mobile apps share account state with the web app — sign in, and your conversation history is available across devices. Mobile-specific features include:

- **Voice input** — dictate prompts instead of typing
- **Image capture** — take a photo to upload as a source
- **Push notifications** — for long-running thinking-mode requests

Mobile feature parity with the web is generally close, with new features arriving on web first.

Browser Extensions

A few browser extensions integrate DeepSeek into the browsing experience — selection-to-summary, side-panel chat, page-content question answering. These are typically third-party rather than DeepSeek-official. Quality varies; check reviews before installing anything that gets access to your browsing.

Privacy and Data on the Web App

The web app stores your conversations on DeepSeek's servers. This has implications for what kinds of material are appropriate to put into it.

A few practical principles:

Don't put confidential business material in the web app. Internal strategic documents, customer data, intellectual property — none of it belongs in a free chat product hosted by any provider, DeepSeek included. For sensitive work, use the API with appropriate contractual protections, or self-host.

Don't put personal sensitive material in the web app. Medical history, financial details, legal questions, anything you wouldn't post on a public forum.

Be aware of the regional data context. DeepSeek operates under Chinese data regulations. Depending on your jurisdiction and compliance requirements, this may or may not be acceptable for specific workloads.

We cover the data-and-privacy story in detail in Chapter 31, including the China question and what it means for serious users.

When the Web App Is the Right Tool

The web app is excellent for:

- **Casual questions** — quick research, drafting help, brainstorming
- **Personal writing assistance** — emails, blog posts, creative work
- **Learning DeepSeek** — exploring capabilities before committing to API use
- **Comparison shopping** — testing the same prompts against ChatGPT, Claude, and DeepSeek to see which fits your workflow
- **Mobile use** — having a capable AI in your pocket without paying for a subscription

The web app is the wrong tool for:

- **Confidential or regulated work** — use the API or self-host instead
- **Production applications** — the web app isn't an API; for embedding into your own products, you need the actual API
- **Workloads requiring custom system prompts** — the web app doesn't expose system prompt control; the API does
- **Workloads requiring fine-tuned model behavior** — same; API or self-hosted only
- **Scripting or automation** — the web app is for humans; API for machines

A Practical First Hour

If you're new to DeepSeek and want to get a real sense of it, here's a productive first hour:

1. **Sign up** at chat.deepseek.com.
2. **Ask a question you know the answer to** — something specific to your domain. See whether DeepSeek answers correctly. Calibrate trust.
3. **Ask the same question with thinking mode on.** Note the differences in approach and answer quality.
4. **Upload a PDF** of something you're familiar with. Ask questions about it. Notice the citation behavior.

5. **Try web search** for a question about something that happened recently. Click the citations to verify.
6. **Compare a complex prompt** between DeepSeek and whichever AI you've been using. Run the same prompt in both. Note where each is stronger.
7. **Try a few politically-sensitive prompts** (just to see) — questions about historical events, political figures. See where the boundaries are.

After that hour you'll have a calibrated sense of what DeepSeek does well, where its limitations are, and whether the web app is enough or you need the API.

The next chapter is for users who've decided they want to integrate DeepSeek into their own applications. We cover getting started with the API — keys, pricing, your first call.

Chapter 4: Getting Started with the API

The web app is for humans. The API is for machines.

Almost every application of DeepSeek that goes beyond casual chat eventually involves the API — a commercial product or internal tool calling DeepSeek programmatically, folding responses into a larger workflow. The API is what makes DeepSeek's economics matter at scale. A team paying $300 a month for OpenAI usage might pay $30 for DeepSeek, with the difference compounding across the year.

This chapter walks through getting the API set up, understanding the pricing model, and making your first call. It assumes some technical background — if you're reading the API chapters, you can read code. The examples use Python and curl; the API is HTTP, so anything that can make HTTP calls can use it.

Creating an API Key

Go to **api-docs.deepseek.com** (or the platform.deepseek.com console — both link to the same place). You'll need a DeepSeek account; the same one that works for chat.deepseek.com works here, or you can create a separate account specifically for API use.

In the console:

1. Navigate to **API Keys** (under Account or Settings)
2. Click **Create new key**
3. Give the key a name (something like `production`, `local-dev`, or `evals` — anything that helps you remember what this key is for)
4. Click create
5. **Copy the key immediately** — DeepSeek (like every responsible API provider) shows the key value only once at creation time. If you lose it, you'll need to delete the key and create a new one.

Store the key somewhere secure. Environment variables are the standard approach — never hardcode it in source code, never commit it to version control.

```
export DEEPSEEK_API_KEY="sk-..."
```

For production use, store the key in your secrets manager of choice (AWS Secrets Manager, GCP Secret Manager, HashiCorp Vault, your platform's equivalent). For local development, a `.env` file outside version control is fine.

Adding Credit

DeepSeek's API is paid; you need credit on your account before calls work. The console has a **Top up** or **Billing** section where you can add credit via standard payment methods (credit card, in some regions Alipay or other regional methods).

A few practical notes:

Start small. \$5–\$10 of credit is enough to do substantial development and testing. You don't need to load \$500 to begin.

Set spending alerts. The console has notification settings for low-balance alerts and spending thresholds. Configure these. Surprise bills are unpleasant; threshold alerts make them rare.

Set hard caps if available. Some account configurations support hard spending caps that pause API access at a threshold. Useful for development environments where a buggy script could burn through credit.

The Pricing Model

DeepSeek's API charges per **token** — the unit of text the model processes. Roughly: one token is three-quarters of a word in English, less in other languages.

Pricing has four components:

Input tokens — what you send to the model in your prompt. Charged per million tokens.

Output tokens — what the model generates in response. Typically charged at a higher rate than input.

Cache hit tokens — input tokens that match a cached prefix from a previous request, charged at a deeply discounted rate (often 80–90% off the cache-miss price). Caching is one of the highest-leverage cost optimizations DeepSeek offers; we cover it in detail in Chapter 12.

Cache miss tokens — input tokens that don't match a cached prefix, charged at the standard input rate.

As of early 2026, with the discount running through May:

- **DeepSeek-V4-Pro**: roughly $0.43/M input (cache miss), $0.0036/M input (cache hit), $0.87/M output
- **DeepSeek-V4-Flash**: roughly $0.14/M input (cache miss), $0.0028/M input (cache hit), $0.28/M output
- Legacy V3.2/R1 endpoints: similar order of magnitude, slightly different ratios

For comparison, GPT-5.5 lands around $5–$15 per million input tokens and $25–$45 per million output, depending on the variant. The cost gap is an order of magnitude or more.

The discount through May 2026 has been a headline feature of V4's launch. After it expires, prices will adjust upward but should still sit well below Western alternatives. Always check current pricing before doing serious cost projections.

Your First API Call

Time to call the API. DeepSeek's API is OpenAI-compatible — it accepts requests in the same shape as OpenAI's chat completions API. You can use any OpenAI-compatible client library, including OpenAI's official Python SDK, just by pointing it at DeepSeek's base URL.

With curl

The bare-metal version. If this works, your account and key are functional:

```
curl https://api.deepseek.com/v1/chat/completions \
  -H "Authorization: Bearer $DEEPSEEK_API_KEY" \
  -H "Content-Type: application/json" \
  -d '{
    "model": "deepseek-v4-flash",
    "messages": [
      {"role": "user", "content": "What is the capital of France?"}
    ]
  }'
```

You should get back a JSON response with the model's answer in `choices[0].message.content`.

With Python (OpenAI SDK)

Install the OpenAI client (`pip install openai`), then:

```
from openai import OpenAI
import os

client = OpenAI(
    api_key=os.environ["DEEPSEEK_API_KEY"],
    base_url="https://api.deepseek.com/v1",
)

response = client.chat.completions.create(
    model="deepseek-v4-flash",
    messages=[
        {"role": "user", "content": "What is the capital of France?"}
    ],
)

print(response.choices[0].message.content)
```

Same shape as a normal OpenAI call. The only difference is the `base_url` and the model name.

With Node.js

```
import OpenAI from 'openai';

const client = new OpenAI({
  apiKey: process.env.DEEPSEEK_API_KEY,
  baseURL: 'https://api.deepseek.com/v1',
});

const response = await client.chat.completions.create({
  model: 'deepseek-v4-flash',
  messages: [
    { role: 'user', content: 'What is the capital of France?' },
  ],
});

console.log(response.choices[0].message.content);
```

Same pattern. Same compatibility.

Reading the Response

Every chat completion response includes the answer plus metadata. The most important fields:

- **`choices[0].message.content`** — the actual response text
- **`choices[0].message.role`** — always `"assistant"` for chat completions
- **`choices[0].finish_reason`** — why generation stopped (`"stop"` for normal completion, `"length"` if it hit the max output token limit, `"tool_calls"` if a tool was called, etc.)
- **`usage.prompt_tokens`** — how many input tokens the call consumed
- **`usage.completion_tokens`** — how many output tokens the response generated
- **`usage.total_tokens`** — sum of input and output
- **`usage.prompt_tokens_details.cached_tokens`** — how many of the input tokens were cache hits (this matters for cost; see Chapter 12)

Multiplying token counts by per-token costs gives you the actual dollar cost of each call. This is useful for cost estimation and for spot-checking that caching is working as expected.

Model Names

The model parameter accepts these names as of early 2026:

- **`deepseek-v4-pro`** — current flagship
- **`deepseek-v4-flash`** — current cost-efficient variant
- **`deepseek-chat`** — legacy alias, currently routes to V3.2
- **`deepseek-reasoner`** — legacy alias, currently routes to R1
- Specific version pins (e.g., `deepseek-v3.2`, `deepseek-r1`, `deepseek-v4-pro-2026-04-24`) when you need reproducibility

For new code, use the explicit version names rather than the legacy aliases. The aliases will eventually be deprecated; explicit names won't change behavior under you.

We'll cover model selection in code (including dynamic switching, routing strategies, and fallback patterns) in Chapter 11.

Understanding Tokens

If you're new to thinking in tokens, two practical anchors:

Token ≈ word, roughly. For English text, a token is typically about 0.75 of a word. So a 1,000-word essay is roughly 1,300 tokens. A 10,000-word document is roughly 13,000 tokens. The 1M-token context window holds approximately 750,000 words — about 1,500 pages of typical book text.

Token ≠ character. A short word like "the" is one token. A long word like "internationalization" might be three or four. Code, especially syntax-heavy languages, often tokenizes inefficiently — Python decorators and TypeScript type annotations both consume more tokens than their character count suggests.

For precise estimation, OpenAI's tokenizer libraries (which work for DeepSeek too, since tokenization is broadly compatible) let you count tokens locally before sending requests. The `tiktoken` library (`pip install tiktoken`) is the standard choice.

Streaming Responses

For interactive applications (chatbots, real-time UIs), you'll want streaming responses — partial output delivered as the model generates it, instead of waiting for the full response.

Add `stream=True` to your request:

```
stream = client.chat.completions.create(
    model="deepseek-v4-flash",
    messages=[
        {"role": "user", "content": "Write a haiku about the ocean."}
    ],
    stream=True,
)

for chunk in stream:
    delta = chunk.choices[0].delta.content
    if delta:
        print(delta, end="", flush=True)
```

The response is a stream of `ChatCompletionChunk` objects, each containing a delta (the next chunk of generated text). Concatenating the deltas reconstructs the full response.

Streaming costs the same as non-streaming. The benefit is purely UX — users see progress instead of waiting silently.

Error Handling

The API can return errors for various reasons:

- **401 Unauthorized** — API key is invalid, missing, or revoked
- **402 Payment Required** — your account is out of credit
- **429 Too Many Requests** — you've hit a rate limit; back off and retry
- **500/502/503** — server-side issue; retry with exponential backoff
- **400 Bad Request** — your request shape was invalid; check the error message for specifics

Every production API call should be wrapped in retry logic for transient errors (5xx and 429) and clean error reporting for permanent errors (4xx other than 429). Most OpenAI-compatible client libraries include retry helpers; configure them rather than rolling your own.

We'll cover authentication, rate limits, and operational concerns in detail in Chapter 10. For now, the basics above are enough for real work.

What You've Just Done

If you've followed along this far, you have:

- A DeepSeek API key
- Some credit on your account
- A working API call returning real responses
- A basic understanding of pricing, models, and token counting

That's the foundation. The rest of the book is about doing more with what you've set up — handling conversations skillfully, using thinking mode, calling tools, structuring outputs, optimizing costs, and eventually building real applications.

The next chapter steps back and walks through your first real conversations and first real API workflows in both surfaces — chat.deepseek.com and the API — so you have a feel for the shape of the work before we go deeper.

Chapter 5: Your First Conversations and First Calls

Theory is good. Practice is better. This chapter takes you through real first uses of DeepSeek across both surfaces — chat.deepseek.com for the web app side and the API for the developer side. By the end you'll have a feel for what working with DeepSeek looks like across the kinds of tasks you'll keep coming back to.

If you're a casual user, the web app sections are where to focus. If you're a developer evaluating DeepSeek for production use, the API sections are where the real work starts. Most readers will get value from both.

Web App: Ten First Prompts

Sign in to chat.deepseek.com. Try these prompts in order. Each illustrates something specific about how DeepSeek handles requests, and the combination gives you a calibrated feel for the tool.

1. Plain explanation

> *"Explain in plain language what makes a Mixture-of-Experts AI model architecturally different from a standard transformer."*

Notice the structure of the explanation. DeepSeek tends toward clear, direct prose with reasonable depth. Compare it mentally to what ChatGPT or Claude would produce on the same prompt — same general accuracy, slightly different rhetorical voice.

2. Code generation

> *"Write a Python function that takes a list of email addresses and returns only the ones that have valid syntax. Use Python's built-in libraries; no third-party dependencies."*

Read the code carefully. DeepSeek is strong on coding. Note whether it explains its choices, how it handles edge cases (what about plus-addressing? what about uppercase letters in the domain?), and whether the code is correct. Run it if you want.

3. Comparison and analysis

> *"Compare the trade-offs between SQL and document databases for a startup building a social network. Be specific about which workloads favor each."*

Notice the structure of the answer. Comparison prompts are where you can see whether DeepSeek understands the topic or is doing surface-level pattern matching. For technical comparisons in software, DeepSeek is generally strong.

4. Writing assistance

> *"Rewrite this email to be more direct: [paste a wordy email of yours, real or invented]."*

Note how DeepSeek handles voice and tone. For utilitarian writing tasks (rewriting, summarizing, translating), the model is competent — competitive with the major Western alternatives.

5. Math problem (no thinking mode)

> *"What's the probability of getting at least three sixes when rolling six six-sided dice?"*

Note the answer. With thinking mode off, DeepSeek will produce an answer but may take some shortcuts.

6. Same math problem (with thinking mode on)

Toggle thinking mode on, ask the same question again.

> *"What's the probability of getting at least three sixes when rolling six six-sided dice?"*

Notice the difference. Thinking mode walks through the binomial distribution explicitly, sets up the problem, computes step by step, and arrives at an answer. The answer is far more likely to be correct (and far more legible if you want to verify it). Compare the two responses to see how much the reasoning step matters.

7. Source-grounded query (web search)

Toggle web search on.

"What was OpenAI's most recent product announcement, and what was the AI community's reaction?"

Notice the citations. Click through to verify. Compare to what the same prompt produces on Perplexity or ChatGPT with search enabled.

8. Document analysis (file upload)

Upload a PDF of any moderately substantial document — a research paper, an article you saved, a report. Then ask:

"What's the central argument of this document, and what's the strongest piece of evidence offered for it?"

Notice how DeepSeek summarizes and how it handles the source content. For document analysis, DeepSeek is competitive with Claude and Gemini.

9. Politically-loaded prompt

"What happened during the 1989 Tiananmen Square protests?"

Notice the response. You'll likely get a refusal or a heavily-scripted non-answer. This is a consistent feature of using DeepSeek's hosted services. We cover the implications in Chapter 31.

10. Meta question

"You're a helpful AI assistant. What kinds of questions are you not great at answering? Be specific."

Notice the self-description. AI models often produce useful self-assessments when asked directly. DeepSeek's tend to be reasonable and reasonably honest about the model's limits.

After these ten prompts, you'll have a working sense of what DeepSeek does well and where its constraints are. Most users find the strengths align with their daily use, with the political-content limitations being the main thing to watch.

API: A Working Chatbot in 30 Lines

For developers, the equivalent first hour is building something. Here's a working command-line chatbot in about 30 lines of Python that demonstrates the core API patterns:

```
import os
from openai import OpenAI

client = OpenAI(
    api_key=os.environ["DEEPSEEK_API_KEY"],
    base_url="https://api.deepseek.com/v1",
)

messages = [
    {"role": "system", "content": "You are a helpful AI assistant. Be direct and
 concise."}
]

print("Chatbot ready. Type 'quit' to exit.\n")

while True:
    user_input = input("you: ").strip()
    if user_input.lower() in {"quit", "exit"}:
        break
    if not user_input:
        continue

    messages.append({"role": "user", "content": user_input})

    response = client.chat.completions.create(
        model="deepseek-v4-flash",
        messages=messages,
    )
    assistant_message = response.choices[0].message
    messages.append(assistant_message)

    print(f"\nai: {assistant_message.content}\n")
```

Run it. Have a conversation. Note that the bot maintains context — your second message can reference what was said in your first, because the entire `messages` history is being sent with each call.

A few things to notice:

Context grows. Every turn, the message history grows. Eventually you'll hit the context window or your token bill will get uncomfortable. For production chatbots, you'll need a strategy for trimming old messages or summarizing them. We cover context management in Chapter 6.

Latency is real. Each response takes a moment — roughly 1–3 seconds for short responses on V4 Flash, longer for V4 Pro or thinking-mode calls. For interactive applications, streaming makes this feel faster.

Costs are tiny. Run this for an hour of casual conversation and check your DeepSeek billing. You'll be surprised how little it cost — a few cents at most.

API: A Streaming Version

Update the request to stream:

```
print(f"\nai: ", end="", flush=True)
stream = client.chat.completions.create(
    model="deepseek-v4-flash",
    messages=messages,
    stream=True,
)
content_parts = []
for chunk in stream:
    delta = chunk.choices[0].delta.content
    if delta:
        content_parts.append(delta)
        print(delta, end="", flush=True)
print("\n")

assistant_message = {"role": "assistant", "content": "".join(content_parts)}
messages.append(assistant_message)
```

Now you see the response unfold in real time. Same cost, much better user experience for interactive use. Most production chatbots stream by default.

API: Thinking Mode

To enable thinking mode in the API, pass the `thinking` field via `extra_body` on the OpenAI SDK (set `{"type": "enabled"}` to turn it on, `{"type": "disabled"}` to turn it off):

```
response = client.chat.completions.create(
    model="deepseek-v4-flash",
    messages=[
        {"role": "user", "content": "What's the probability of getting at least
three sixes when rolling six six-sided dice?"}
    ],
    extra_body={"thinking": {"type": "enabled"}},
)
```

The response includes a `reasoning_content` field with the chain-of-thought alongside the normal `content` with the final answer. One gotcha for multi-turn conversations: strip `reasoning_content` from prior assistant messages before sending them back as input — leaving it in causes a 400 error. The exception is after a tool call, where `reasoning_content` MUST be included in the follow-up.

For workloads where reasoning matters, thinking mode is one of DeepSeek's strongest features. Chapter 16 covers when to use it.

API: A Simple Summarizer

A more practical first project — a function that takes any text input and returns a structured summary:

```
import os
from openai import OpenAI

client = OpenAI(
    api_key=os.environ["DEEPSEEK_API_KEY"],
    base_url="https://api.deepseek.com/v1",
)

def summarize(text: str, max_words: int = 200) -> str:
    response = client.chat.completions.create(
        model="deepseek-v4-flash",
        messages=[
            {
```

```
                "role": "system",
                "content": (
                    "You produce tight, structured summaries of text. "
                    "Format: a one-sentence headline, then 3-5 bullet points "
                    "of the most important specifics, then a one-sentence takeaw
ay."
                ),
            },
            {
                "role": "user",
                "content": f"Summarize this in approximately {max_words} words:\
n\n{text}",
            },
        ],
    )
    return response.choices[0].message.content

# Usage
with open("article.txt") as f:
    article = f.read()
print(summarize(article))
```

This pattern — a system prompt that establishes behavior, a user prompt that provides the input — is the template most production AI workloads follow. The variations are about prompt design, model selection, output structure, and cost management. The basic shape stays.

What You've Just Done

If you've worked through the chapter:

Web app side: You've used DeepSeek for a range of tasks (explanation, code, analysis, writing, math, web research, document analysis), seen where thinking mode helps, encountered the political-content boundaries, and gotten a calibrated feel for what the tool does well.

API side: You've made working API calls, built a basic chatbot with conversation memory, tried streaming responses, enabled thinking mode in code, and built a usable summarization function.

That's enough to do real work. From here:

- For deeper web-app skills, continue with Part II (Chapters 6–9)

- For deeper API skills, jump to Part III (Chapters 10–15)
- For self-hosting, head to Part V (Chapters 19–22)
- For specific use cases, see Part VI (Chapters 23–26)

The next chapter (the start of Part II) goes deeper on the web app: managing context, when to start new chats, reading confidence, verifying answers — the conversational craft that separates power users from casual ones.

That closes Part I. You now have the foundation: what DeepSeek is, the model lineup, the web app and API surfaces, your first hands-on experience with both. The rest of the book goes deeper across each of those threads.

Part II: The Web Chat Experience

"The simplest tool, used well, beats the most powerful tool used carelessly."

— A working principle

Chapter 6: Conversations and Context

The web app looks like a chat window. You type, DeepSeek answers, you type again. It's familiar enough that most users never think about what's happening underneath. That's fine for casual use, but the moment you want serious work out of DeepSeek — long research sessions, complex multi-turn projects, careful analysis — understanding how the conversation works matters.

This chapter covers the things every web-app user benefits from understanding: how DeepSeek manages conversation context, when to start a new chat versus continue the current one, how to read the model's confidence and verify what it tells you, and what the 1M-token context window fits.

How Context Works

When you ask DeepSeek a question in the middle of a conversation, the model isn't reading just your latest message. It's reading the full history of the conversation up to that point — every prompt you've typed, every response the model has generated. That history is the **context**, and it's what lets DeepSeek answer follow-up questions, refer back to earlier topics, and build on what's been said.

A few practical implications:

Earlier messages affect later answers. If you asked DeepSeek to roleplay as a particular character three messages ago, it'll still be roleplaying now. If you set a constraint ("respond only in bullet points"), that constraint persists for the rest of the conversation. If you fed in a document in your second message, DeepSeek can still reference it in your tenth.

Each turn re-sends everything. Every message you send includes the entire conversation history. This is why long conversations get slightly slower over time — the model is processing more input on each turn.

Context costs add up. On the free web app this shows up as latency, not a bill. On the API (where you pay per token) it shows up directly. We cover the API economics in Part III.

The model can't see everything forever. Context windows are large (1M tokens for V4) but finite. Push past the limit and the earliest messages drop out of the model's view. The practical implications are below.

The 1M-Token Context Window

DeepSeek-V4's context window of 1 million tokens is enormous by historical standards. To put it in concrete terms:

- A typical novel is around 100,000 words ≈ 130,000 tokens — easily fits, with room for substantial conversation about it
- A 200-page nonfiction book is around 75,000 words ≈ 100,000 tokens — comfortable
- The complete works of Shakespeare are around 900,000 words — most of it would fit
- A 1,500-page legal code might exceed the limit, but most things don't

For interactive web-app conversations, hitting the context limit is almost impossible in practice. You'd need a multi-hour, multi-document, deeply technical conversation before the limit becomes the binding constraint.

Where the limit matters:

- **Long documents with deep questioning** — uploading a 600-page report and having a 50-message conversation about every section
- **Multi-document analysis** — uploading a stack of papers and asking comparative questions across all of them
- **Very long-running research projects** — coming back to the same conversation for weeks of work

Most users will hit other limits — their own attention span, the model's ability to maintain coherence over very long sessions — before the context window.

When to Start a New Chat vs. Continue

This is the most consequential workflow decision in the web app. Get it right and the conversation works for you. Get it wrong and you'll fight the tool.

Start a new chat when:

- The topic shifts. You've been talking about marketing strategy and now you want help with a Python script. Don't mix them.
- The previous answer set you on a wrong path. If the conversation has accumulated wrong assumptions, those assumptions persist. Sometimes the cleanest reset is a new chat.
- You want to start fresh on something complex. New conversations get the model's full attention without the drag of accumulated context.
- The conversation is getting circular. If you're hearing the same points repeated or the model seems "stuck," start over with a tighter framing.
- You want to share or save a clean conversation. If the work-in-progress chat is messy, do the cleanup in a fresh one once you know what you want.

Continue the current chat when:

- You're refining or going deeper on the same topic. Follow-ups, clarifications, more depth — these benefit from accumulated context.
- You want the model to reference earlier material. "Compare what you just said about X to the Y you mentioned earlier" only works in the same conversation.
- You've established a useful framing or constraint. If you spent ten messages getting the model to write in a particular voice, don't abandon that work.
- You've uploaded documents that you want to keep using. The uploaded files persist in the conversation.

A common mistake is keeping one mega-conversation that handles everything. It works for a while, then accumulates so much unrelated context that new questions get diluted. A better pattern: short-to-medium conversations focused on specific tasks, with new chats started liberally.

Reading Confidence and Verifying Answers

DeepSeek, like every large language model, can be wrong. Sometimes confidently wrong. Reading the signals that distinguish "the model knows this" from "the model is filling in plausibly" is one of the most important user skills.

A few patterns to watch for:

Hedging language usually means low confidence. Phrases like "I believe," "as far as I know," "this may have changed," "approximately," "around" — these are often signals that the model is uncertain about the specific claim.

Specific numbers and dates need verification. The model is statistically likely to get a number close to right, but exact figures (specific revenue numbers, dates, addresses, phone numbers) are where hallucination shows up most. If the answer matters, verify.

Confident-sounding citations need verification too. DeepSeek sometimes cites papers, books, or articles that don't exist, or attributes claims to real sources that didn't make them. Don't quote a citation without confirming it.

Math and code are testable. The advantage of these domains is that you can check the answer. Run the code. Verify the calculation. The model will sometimes produce confidently wrong answers in both, and you'll catch it because the code throws an error or the math doesn't add up.

For controversial or contested claims, ask the opposite. If the model tells you confidently that X is true, ask "what's the strongest case that X is false?" If the model produces a thoughtful counter-argument, you've learned the issue is contested. If it stumbles, the original confidence was overstated.

The discipline this requires is your own attention. The model won't tell you when it's guessing. You learn to spot it through experience.

Switching Models Mid-Conversation

The web app lets you switch the active model partway through a conversation. The next message and beyond will use the new model; the conversation history stays intact.

Common patterns:

Start with Flash, escalate to Pro. Use the cheaper, faster model for exploratory back-and-forth. When you reach a question that needs the more capable model, switch.

Toggle thinking mode for hard questions. You don't need to switch models — just toggle thinking mode on for the message that requires it.

Drop back to Flash when the heavy lifting is done. If you spent a few messages with Pro on the analytical work and now you're polishing prose, Flash is fine.

The switch isn't free — the new model has to read the entire conversation history before responding, which costs a bit more compute. But it's usually a worthwhile trade.

Managing Long Conversations

For conversations that genuinely run long — a research session that spans an afternoon, a multi-document analysis, a complex creative project — a few practices help:

Periodically summarize what you've established. Ask the model to "summarize what we've concluded so far." Save the summary somewhere outside the chat (a notes file, your own document). That gives you a checkpoint you can come back to if the conversation derails.

Branch off for sub-tasks. When you need to dig into a specific sub-problem, start a new chat with the relevant context (paste in the summary from the parent conversation). When you're done, bring the result back to the main conversation.

Watch for drift. Long conversations sometimes accumulate small misunderstandings that compound. If answers start feeling off, scroll back through the history and check whether something earlier set the model on a wrong path.

Don't be afraid to restart. A conversation that's not working is a sunk cost. Restart with what you've learned and a tighter framing. The new conversation often goes faster than continuing the old one.

Context-Specific Quirks Worth Knowing

A few specific behaviors of DeepSeek's web-app context handling worth noting:

Files uploaded earlier persist. A PDF you added in your second message is still available in your tenth. You don't need to re-upload.

Web search results don't persist as searchable context. When you run a web-search-grounded query, the model uses the search results to answer that specific message. Subsequent questions don't automatically search; they treat the previous answer as known content.

Thinking-mode reasoning isn't always preserved. When the model produces explicit chain-of-thought reasoning in thinking mode, the reasoning may or may not be visible to subsequent turns. The final answer is preserved; the work-in-progress reasoning sometimes isn't.

Code interpreter outputs persist as text. If the model executed code and produced output (a chart, a table, a calculation result), that output is part of the conversation context going forward. You can ask follow-up questions about it.

A Working Pattern: The Project-Centric Conversation

For users doing serious work in the web app, a useful pattern is the **project-centric conversation** — one chat focused on a specific outcome (a paper, a product launch, a research project), with all the work happening in that single thread.

The discipline:

1. Start the conversation with a clear framing: "I'm working on [specific project]. The goal is [specific outcome]. The audience is [specific audience]. Constraints: [specific constraints]."
2. Upload any source documents up front so they're available throughout.
3. Reference what's been established as you go: "Building on the structure we sketched, what's the strongest opening?"
4. Periodically pause to summarize: "Let's recap what we've decided about X."
5. When you finish a sub-task, save the relevant output (paste into your real document, save the result file) before moving on.
6. When you genuinely shift topics, start a new conversation.

Used this way, the web app becomes a working environment, not just a Q&A interface. The conversation is the workspace.

Chapter 7: Thinking Mode

In November 2024, OpenAI released the o1 family — language models trained to produce explicit chain-of-thought reasoning before their final answers. The internal monologue, made visible. Within weeks the industry was racing to ship reasoning models. DeepSeek's R1 (January 2025) was an early entrant and set a new bar: reasoning quality comparable to o1 at a fraction of the cost, with the weights released under MIT license.

By 2026, **thinking mode is a per-call feature in V4** rather than a separate model line. Toggle it on when you want the model to think; toggle it off when you don't. The decision matters more than most users realize: thinking mode dramatically improves answers on certain tasks and is pure waste on others.

This chapter covers what thinking mode is, when to use it, when to skip it, and how to read the reasoning the model produces.

What Thinking Mode Is

When you submit a prompt with thinking mode enabled, DeepSeek generates a response in two parts:

The reasoning — a chain of thought where the model works through the problem. It poses sub-questions, considers alternatives, checks its own work, sometimes catches its own mistakes. This appears in a designated section of the response (in the web app, often visually distinct; in the API, often in a separate field).

The final answer — the model's synthesized response, informed by the reasoning above.

The reasoning is real, in the sense that the model is doing computational work to produce it, and the final answer typically reflects that work. It's not just for show. For problems that benefit from step-by-step thinking, the quality improvement is genuine.

How Thinking Mode Differs from Regular Responses

Without thinking mode, DeepSeek produces an answer directly. Behind the scenes the model does some implicit reasoning — neural networks always do — but it's not exposed and is typically less thorough.

With thinking mode, the model is trained to spend computational tokens on explicit reasoning before the final answer. The token cost is higher (more output tokens), the latency is higher (more computation), and for the right problems, the answer quality is meaningfully better.

The improvements are biggest in:

Multi-step reasoning — problems where you need to do A, then B based on A, then C based on B. The explicit trace forces the model to do the steps in order.

Mathematical work — algebra, probability, statistics, calculus. The reasoning step makes the model work through the problem the way a careful human would.

Logical analysis — formal logic, conditional reasoning, "if X then Y, but if Z then..." chains. Thinking mode handles these far better than direct response.

Code that has to work — for non-trivial code, thinking mode helps the model consider edge cases, plan the structure, and catch errors before producing the final code.

Complex planning — multi-stakeholder strategies, sequenced approaches, anything where the answer is a plan rather than a fact.

The improvements are smaller (or zero) in:

Simple factual questions — "What's the capital of France?" doesn't benefit from thinking mode. The model knows.

Pure stylistic writing tasks — rewriting an email, generating poetry, polishing prose. These reward feel and craft, not deliberate reasoning.

Open-ended brainstorming — when you want quantity and variety, thinking mode narrows the output by working through one path more thoroughly.

Creative roleplay or fiction — the explicit reasoning often interferes with the creative voice.

When to Use Thinking Mode

A practical rule of thumb: **enable thinking mode when you'd want a careful human to think before answering.**

Some specific cues:

- The answer requires multiple steps you can't shortcut
- A wrong answer would be costly (to your work or to your trust in the tool)
- The domain rewards explicit reasoning (math, formal logic, careful coding)
- You want to see the work, not just the answer
- The answer space is large and you want the model to consider alternatives

And some cues for when to skip thinking mode:

- The answer is a single fact or short retrieval
- The task is pure pattern generation (writing, coding boilerplate, generating examples)
- You're doing fast iterative work and the thinking-mode latency would be disruptive
- The model has confidently produced good answers on similar problems before

You can mix freely within a single conversation. Toggle thinking on for the messages that need it, off for the ones that don't. The model handles the per-message switch cleanly.

The Tradeoffs

Thinking mode is more expensive in three ways:

Token cost. Reasoning tokens are output tokens, and output tokens cost more than input tokens. A thinking-mode response of 2,000 reasoning tokens plus 500 final-answer tokens costs more than a direct 500-token response.

Latency. The reasoning takes time to generate. A thinking-mode call that would have taken 2 seconds direct might take 8–15 seconds. For interactive workflows where you're waiting on each response, this adds up.

Distraction. Reading a long reasoning trace before reaching the final answer can be distracting in casual contexts. The model shows its work; you have to skim or skip past it.

For high-volume API workloads, these costs scale. A production system that uses thinking mode on every call, regardless of need, will spend substantially more than one that uses it selectively. We cover the cost-optimization patterns in Chapter 16.

For the web app, the costs are mostly in your time and attention rather than dollars. Use thinking mode liberally when it helps; skip it when you'd rather have a fast direct answer.

Reading the Reasoning

When thinking mode produces a reasoning trace, what do you do with it?

A few practical uses:

Verify the model's logic. If the reasoning is sound, you can trust the final answer more confidently. If you can spot errors in it, you've caught a problem before it propagates.

Learn from the model's approach. For problems in domains you're learning, the trace is often educational. You see how a competent reasoner might approach the same problem.

Catch confident mistakes. Sometimes the reasoning shows exactly where the model went wrong — a misread of the question, an arithmetic mistake, an unjustified assumption. The trace makes the failure debuggable.

Pull useful intermediate results. Sometimes the most valuable part of the response isn't the final answer — it's a clarifying sub-conclusion the model reached along the way.

The reasoning isn't a perfect window into what the model "really thinks." Language models don't have private mental states; the trace is itself just generated text, calibrated through training to produce useful intermediate work. But it's useful generated text, and treating it as a working artifact rather than a magic ledger is the right frame.

Math, Code, and Multi-Step Problems: Worked Examples

Three short worked examples to show what thinking mode does in practice.

Example 1: A probability question.

Without thinking mode:

> *Q: If I roll three fair six-sided dice, what's the probability that the sum is exactly 10? A (no thinking): The probability is approximately 1/8 or 12.5%.*

With thinking mode:

> *Q: Same question. A (with thinking): [The reasoning explicitly enumerates the ways to sum to 10 with three dice — (4,4,2), (4,3,3), (5,4,1), etc. — counts the permutations, divides by 216 total outcomes, arrives at 27/216 = 12.5%.] Final: The probability is exactly 27/216, which simplifies to 1/8 (12.5%).*

Both answers happen to be correct, but the thinking-mode answer is verifiable — you can check the enumeration. The non-thinking answer is right but you have to trust it.

Example 2: A code task.

> *Q: Write a Python function that takes a string and returns true if all characters are unique, false otherwise. Do it in O(n) time without using any built-in collection types.*

Without thinking mode, the model might produce something using a `set` (a built-in collection — violates the constraint). With thinking mode, it's more likely to weigh the constraints carefully, settle on a bit-vector approach, and produce code that meets the requirements.

Example 3: A planning question.

> *Q: I'm migrating a 50-engineer team off Postgres onto a sharded MySQL setup. What's the safest 6-month rollout plan that keeps writes consistent during the transition?*

Without thinking mode: a competent but generic answer covering common migration patterns. With thinking mode: the model works through the dual-write problem, the read-cutover decisions, the rollback considerations, and the team-coordination dependencies, arriving at a more specifically reasoned plan.

For all three, thinking mode produces a genuinely better answer. For "what's the capital of France," it doesn't.

Thinking Mode in Long Conversations

A few practical notes on using thinking mode across multi-turn conversations:

The reasoning isn't always preserved across turns. As noted in Chapter 6, the model's chain-of-thought from a previous response may or may not be visible to subsequent turns. Don't rely on the model "remembering" its reasoning; if it matters for follow-ups, paraphrase the relevant conclusions.

You can prompt for reasoning explicitly. Even without the toggle: "Walk through your reasoning step by step before giving the final answer." This produces reasoning-mode-like output from non-thinking calls.

Mixed conversations work. Some messages with thinking on, others off, in the same conversation. The model handles the switching cleanly.

What This Means for Daily Use

For most users, thinking mode shifts from "feature you might try" to "discipline you develop." A practical rhythm:

- **Default off** for everyday conversation, brainstorming, casual questions
- **Default on** for math, complex coding, careful analysis, important decisions
- **Toggle deliberately per message** as the conversation moves between modes

The users who get the most out of DeepSeek (and similar reasoning-capable models) are the ones who develop instincts for when reasoning is worth its cost. That instinct comes from use — try thinking mode on a problem you've already tried direct, see whether the answer is better. After a few rounds of comparison you'll know intuitively when to toggle.

The next chapter covers another feature that turns the web app from a chat interface into a working tool: file uploads, code interpreter, and the document-analysis workflows that make DeepSeek useful for real research and analysis.

Chapter 8: File Uploads and Code Interpreter

The chat interface alone is useful but limited. DeepSeek's web app becomes a genuine working tool when you can drop documents into a conversation, upload spreadsheets and ask analytical questions, paste in screenshots and have the model extract what it sees, or have the model execute code to verify a calculation or produce a chart.

This chapter covers the file and code workflows that turn chat.deepseek.com from a smart conversation partner into a working analytical environment.

Supported File Types

As of early 2026, the web app accepts most file types you'd want to work with:

Documents: - **PDF** — the most common upload, well-supported including scanned PDFs (OCR runs automatically) - **DOCX** — Microsoft Word files, including formatting and embedded images - **TXT** — plain text - **CSV** — comma-separated values, useful for tabular data - **MD** — Markdown files

Images: - **PNG, JPG, JPEG, GIF, WEBP** — handled with vision capability + OCR for any embedded text

Spreadsheets: - **XLSX** — Excel files, including multi-sheet workbooks

Code and structured text: - Most source code files (.py, .js, .ts, .go, .rs, .java, etc.) work as plain text uploads

Multiple files per conversation are supported. Size limits vary by file type and may shift between releases — typically tens of megabytes per file is fine; hundreds of megabytes start hitting limits.

Uploading Files

The mechanics are familiar: a paperclip or upload icon in the input area, drag-and-drop into the chat window, or paste a file from your clipboard. Once uploaded, the file is part of the conversation context — DeepSeek can reference it across all subsequent messages until you start a new chat.

A few practical notes:

The first message after upload should reference the file. "Summarize this document," "What are the key findings in this report," "Extract all the dates mentioned." A bare upload with no question doesn't tell the model what you want.

Uploaded files persist in the chat. No need to re-attach a file for follow-up questions. The model can return to the document throughout the conversation.

Multiple files become a corpus. Upload three related papers and you can ask cross-document questions: "How do these three papers differ in their methodology?"

Large files take a moment to process. The model needs to ingest the document before answering. For very large uploads, expect a longer first-response delay.

Document Analysis Workflows

Some patterns that work well for document-heavy use:

The summary-then-drill-down pattern.

1. Upload the document.
2. First prompt: "Summarize this document in three paragraphs. First paragraph: what it is and who it's for. Second: the central claim or finding. Third: structure and where the major sections are."
3. Read the summary. Identify the part that interests you.
4. Drill in: "Go deeper on section [X]. What specific evidence is offered, and how strong is it?"

This pattern takes you from "I have this document" to "I understand what matters in this document" in three or four turns.

The verification pattern.

1. Upload a document.

2. State a claim you've heard about it: "I've read that this paper finds X. Is that what it actually says?"
3. The model checks against the document and tells you whether the claim holds up.

Useful for verifying citations, checking secondhand summaries, and pressure-testing claims from elsewhere.

The cross-document comparison pattern.

1. Upload two or more related documents.
2. Ask comparative questions: "How do these documents agree, and where do they conflict?"
3. Pull specific points: "What does each say about [specific topic]?"

Useful for literature reviews, due diligence, contract negotiations, and any work where multiple sources need to be reconciled.

The structured-extraction pattern.

1. Upload a document containing structured information (a table-heavy report, a contract, a research paper with specific data).
2. Ask for a structured extraction: "Extract every dollar figure mentioned, the context it appears in, and the section number."
3. The model produces a structured response — often as a table — that you can copy out for further analysis.

Useful for taking unstructured documents and producing the structured data you need.

Spreadsheet and CSV Analysis

For tabular data, DeepSeek's combination of file reading and code interpreter (covered below) is particularly powerful:

Quick descriptive analysis. Upload a CSV. Ask: "Describe this dataset. What columns are present, how many rows, any obvious patterns or outliers?"

Targeted questions. "What's the average revenue by region? Are there months that look unusual?"

Visualization on demand. "Plot revenue over time and break out by product category." DeepSeek's code interpreter generates a chart inline.

Data cleaning. "There are some malformed dates in this CSV — find them and propose corrections."

For analytical workflows, this pattern replaces a substantial amount of one-off scripting. You iterate quickly: ask, see, adjust, ask again.

Image Uploads

DeepSeek's vision capability handles images well. Common uses:

Document images. A screenshot, a photo of a printed document, a phone-camera shot of a whiteboard. The model OCRs the text and treats it as readable content.

Charts and diagrams. Upload a chart from a report; ask "What does this show?" or "What's the key takeaway?" The model interprets the visual.

UI screenshots. Helpful for "what does this button do?" or "explain this error message" workflows.

Handwritten notes. OCR for handwriting is hit-and-miss but often surprisingly good, especially for clearly written notes.

Mixed text and images. PDFs with embedded charts work best when the chart is also referenced in surrounding text — the model uses the text context to interpret the visual.

A few limits worth knowing:

Pure visual content (photos, art) is interpreted descriptively. The model can describe what's in an image but lacks the precision a human would bring to visual judgments.

Very small text may not OCR cleanly. A screenshot of fine print (legal text, dense table footnotes) can produce garbled extraction.

Multi-page scanned PDFs are best uploaded as PDFs, not as separate images. The model handles document structure better when the file is a proper PDF.

Code Interpreter

For some queries, DeepSeek can execute code in a sandboxed environment — running calculations, generating charts, processing uploaded data, verifying a piece of work.

The interface for invoking it varies. Sometimes it surfaces inline ("I'll run this for you to check"); sometimes there's an explicit toggle. When the model decides to execute code, you'll see a code block, then output (numbers, a chart, a table).

Common uses:

Calculations the model wants to verify. "What's the present value of $10,000 received annually for 30 years at 4% discount rate?" The model executes the financial math rather than estimating.

Data analysis on uploads. Upload a CSV; ask analytical questions; the model writes and runs Python (typically) to produce the answer.

Chart generation. "Plot this data as a stacked bar chart with months on the x-axis." The model generates the chart and shows it inline.

Code verification. "Here's a function I wrote — run it on these test inputs and tell me if the output is correct."

A few practical notes:

The sandbox is isolated. The execution environment doesn't have your local files or your network. It works with what you've uploaded into the conversation.

Common libraries are pre-installed. Python's standard library, pandas, numpy, matplotlib, and other common analytical libraries are available. Specialized libraries may not be.

Files generated in the sandbox can be downloaded. If the model produces a CSV or a chart image, there's typically a download link.

Sessions are ephemeral. State doesn't persist across conversations. For multi-step analysis, do it in one conversation.

Limits and What to Do When Uploads Fail

A few common failure modes and how to handle them:

"File too large." Reduce the file size – split a long PDF into chapters, downsample large images, export a compressed version. For genuinely huge files, the API may handle them better than the web app, or you may need to chunk the source.

"OCR failed" or garbled text from a scanned document. The scan quality may be too low. Try a higher-resolution scan, or pre-process with a dedicated OCR tool (Tesseract, Adobe Acrobat, Google Document AI) before uploading the cleaner text.

"Format not supported." Less common, but if you have an unusual format, convert to PDF or plain text first.

Images not rendering. Sometimes the upload completes but the model doesn't seem to "see" the image. Try a different format, or describe what's in the image to confirm the model has access.

Generic answers about a document it should have read. The model may not have processed the upload completely, or the file may be too large for its effective context. Split into smaller pieces, or ask more specific questions to verify the model is engaging with the content.

A Practical Walkthrough: Document Triage

To make this concrete: you've received a 60-page report and need to brief your team in 30 minutes. A productive workflow:

Phase 1 – orientation (five minutes).

Upload the PDF. First prompt:

> *"Summarize this report in three paragraphs. First: what it is and who produced it. Second: the central findings. Third: the structure and most important sections."*

Read the summary. You now know the shape of the document.

Phase 2 — depth (15 minutes).

For each major section identified in the summary, ask drill-down questions:

> *"Go deeper on section 3 (the methodology). What approach did the authors take, and what are its strengths and weaknesses?"*
>
> *"Walk me through the key findings in section 4. Quote the most important specific claims."*
>
> *"What recommendations does the report make? Cite the section."*

Save the most important points as you go (paste into a notes file).

Phase 3 — synthesis (five minutes).

> *"Generate a one-page briefing for [specific audience] covering this report. Focus on [specific angle] and skip [things they don't need]."*

Edit the briefing for your team's voice and any specifics. Done.

Phase 4 — verification (five minutes).

For any claim in the briefing that's load-bearing — anything you'll be quoted on — ask:

> *"Quote the exact passage in the report that supports the claim that [X]."*

Confirm the citation. If it doesn't hold up, fix the briefing.

You've gone from receiving the report to a verified briefing in 30 minutes, with the document genuinely engaged rather than skimmed.

What the Web App Doesn't Replace

A few things to keep in mind:

The web app isn't a citation manager. For academic citation tracking, use Zotero, Mendeley, or your institution's tool.

It isn't a replacement for reading the source on high-stakes work. For anything where you'll be on record about what a document says, read the relevant sections yourself.

Code interpreter isn't a replacement for a real development environment. It's good for one-off analysis and verification; it's not where you build production systems.

Spreadsheet analysis through chat isn't a replacement for actual spreadsheet skills. For ongoing analysis you'll do repeatedly, build the spreadsheet or the script properly. Chat is for one-off queries.

Obvious points, but worth stating: the web app is excellent at compressing certain workflows; it isn't a substitute for the underlying tools when those tools are warranted.

The next chapter covers the last of the web app's major features: web search, where DeepSeek pulls in information from the open web to ground responses in current information.

Chapter 9: Web Search Integration

DeepSeek's training data has a cutoff date. Anything after that date — recent product launches, current news, this week's regulatory changes, the most recent earnings reports — isn't in the model. For questions that depend on current information, the model alone isn't enough.

The web app's web search feature solves this by letting DeepSeek fetch information from the open web in real time, then answer your question grounded in what it found. This chapter covers how the feature works, when it's the right tool, and the verification habits that make it useful.

How Web Search Works

When you toggle web search on, here's what happens behind the scenes:

1. DeepSeek receives your prompt.
2. The model formulates one or more search queries based on the prompt.
3. The system retrieves results from the open web.
4. The model reads the retrieved content.
5. The model produces a response, with citations indicating which sources informed which claims.

You see citations as small numbered references in the response (or as hyperlinks, depending on the interface version). Click a citation to go to the source the model used.

A few mechanical notes:

The model decides what to search for. You can guide it with your prompt — be specific about what you want — but the actual query is the model's interpretation.

Multiple sources may be consulted. A complex question can trigger several search queries and synthesize across the results.

The model picks what to use from each result. The full webpage content isn't dumped into your conversation; the model selects what's relevant.

Responses are still subject to the model's quality. Web search supplies information; the model still has to interpret and integrate it.

When DeepSeek Searches vs. Uses Training Data

This matters more than it sounds. With web search enabled, DeepSeek doesn't always actually search — sometimes it answers from training data even when the toggle is on, because the model judged the search unnecessary.

Conversely, with web search disabled, the model uses only what it knows from training. For questions about recent events, that means stale or missing answers.

A few practical implications:

Verify by checking citations. If the response has citations, the model searched. If it doesn't and you asked something requiring current info, it didn't — be skeptical of the answer.

Force a search when you need one. "Search the web and find the most recent..." or "What does the most recent reporting on X say?" pushes the model to use the search.

Recognize when search wasn't needed. A question like "What's the boiling point of water?" doesn't need a web search. The model has it. Toggling search on for stable factual questions just adds latency.

Getting Cited, Sourced Answers

The headline benefit of web search is that responses become **citation-grounded** — every substantive claim points to a source you can verify. This is far more reliable than ungrounded responses for several reasons:

You can verify. Click the citation, read the source, confirm the claim. Non-negotiable for high-stakes use.

The model is more careful when citing. Models trained for citation-grounded responses (DeepSeek's web-search mode included) tend to be more measured in their claims when they know they're being asked to cite. Less hedging, more specifics.

You can follow up at the source. If a citation looks interesting, the source is often more useful than the model's summary. Going to the source is a strength of search-grounded interactions, not a workaround.

The verification habit from Chapter 23 of *NotebookLM: From Sources to Audio Overviews* applies here too: **the citation only proves the source exists. It doesn't prove the source supports the claim.** Click through; verify.

Comparing DeepSeek Search to Alternatives

A few comparative notes for users choosing among AI search tools:

Vs. ChatGPT search. Both work well for general queries. ChatGPT's search has been more aggressive about citation density and source quality filtering; DeepSeek's surfaces fewer sources but with cleaner integration. For most queries, both produce useful results.

Vs. Claude search. Claude's web search has been less mature than DeepSeek's as of early 2026. For workflows where Claude is otherwise the right tool, do search-heavy work in DeepSeek (or Perplexity) and bring the synthesized findings back to Claude.

Vs. Perplexity. Perplexity is purpose-built for search-grounded research and remains the strongest option for "find me information across the web and cite it" workflows. DeepSeek's web search is a feature within a general-purpose chat tool; Perplexity's web search is the product. For research-heavy work, Perplexity is often better. For casual fact-checking and current-events questions in the middle of a normal conversation, DeepSeek's search is fine and saves you from switching tools.

Vs. Google directly. Web-grounded AI tools are good for synthesis across sources; raw Google is better for "I need to find one specific thing" or "I need to evaluate source quality myself before reading the answer." Both have a role.

Real-Time Information and Currency Limits

A few practical notes on what "current" actually means with web-grounded responses:

The web isn't always immediately updated. A breaking news event from an hour ago may not yet be reflected in indexed search results. For truly real-time information (markets, ongoing emergencies), even web search can lag.

Source quality is uneven. Search may surface high-quality reporting, but it may also surface AI-generated content farms, partisan blogs, or outdated articles. Citation quality varies.

Recency bias. The model often weights recent sources more heavily than older but more authoritative ones. Usually right; sometimes wrong (when an old definitive source is being superseded by inferior recent coverage). Spot-check.

Geographic bias. The web is overwhelmingly English-language. For non-English topics, search may surface less authoritative sources than direct queries to language-specific resources would.

Patterns That Work Well

Some specific use cases where web search shines:

Recent events. "What happened at the AI safety summit last week?" — search produces current, citation-grounded coverage.

Product launches and releases. "What did OpenAI announce at their last DevDay?" — search captures the announcements and the community response.

Current pricing or availability. "What's the current pricing for [product]?" — search produces actual current numbers, not training-data approximations.

Recent reviews and comparisons. "What do people think of [tool] in late 2025?" — search produces aggregated current sentiment.

Verifying secondhand claims. "I've heard that [X claim]. Search to find out if this is supported." — produces a sourced check.

Quick fact-checks. "Did [public figure] actually say [quote]?" — produces the source and the context.

Patterns That Don't Work Well

Where web search is less useful:

Stable factual questions. Boiling points, historical dates, mathematical constants. Training data is fine.

Deep analysis of evergreen topics. "Explain the difference between concurrency and parallelism." The model knows; search adds noise.

Anything requiring expertise the open web doesn't cover well. Niche academic topics, specialized professional domains, internal organizational knowledge — search can't help with what isn't on the open web.

Brainstorming and ideation. Search constrains the response toward what already exists. For generating new ideas, search-off is often better.

Highly contested or politicized topics. Search surfaces biased sources. The model attempts neutral synthesis, but the underlying source pool may not support neutrality.

A Practical Walkthrough: Researching a Recent Topic

Suppose you want to understand the current state of a fast-moving topic — say, the consensus on a specific AI policy question. A productive workflow:

Step 1 — broad context (search on).

> *"Search the web. What's the current state of debate around [specific policy question]? Who are the major positions?"*

Read the response. Note the major positions and the cited sources.

Step 2 — drill into one position (search on).

> *"Find the strongest recent argument for [specific position]. Quote and cite."*

Click through the citations to read the source.

Step 3 — counter-argument (search on).

> *"Now find the strongest argument against. Same — quote and cite."*

Again, follow the citations.

Step 4 — recent developments (search on).

"What's happened on this topic in the past month? Any major developments?"

This catches anything time-sensitive.

Step 5 — synthesis (search optional).

"Synthesize what we've covered. What's the strongest case on each side, and where is the actual locus of disagreement?"

Search may or may not be needed; the model can synthesize from what's already in the conversation.

You've done substantive research on a current topic in maybe 20 minutes, with citations to verify everything important.

When Web Search Belongs in the API

For developers building applications, web search is also available via the API — typically through model-specific parameters or tool/function calling integrations. We touch on this in Chapter 13 (tool calling) and Chapter 23 (DeepSeek for developers).

For most production use cases, web search at the API layer is rarely the right architecture. It adds latency, indeterminism (you don't know what sources will be returned), and external dependencies. Production systems that need current information typically integrate search through a controlled provider (Bing, Brave, Tavily, etc.) and feed the results to the model rather than letting the model search opaquely.

For interactive web-app use, the trade-offs flip: convenience matters, you can verify in real time, and the integrated experience beats tool-juggling.

What Closes Part II

That closes Part II. You know how to use the web app well: how to manage conversations, when to use thinking mode, how to work with files and code, when web search is the right tool. These are the day-to-day craft skills for getting real value out of chat.deepseek.com.

Part III shifts to the API — the developer's view of DeepSeek. If you're building applications, the next six chapters are the core of the book. If you're a web-app-only user, skim Part III and pick up again at Part IV (reasoning) or jump ahead to the audience-specific chapters in Part VI.

Either way, what you've learned about conversation, thinking mode, files, and search transfers — the same principles apply on the API side, with more control and more responsibility.

Part III: Building with the API

"What gets measured gets managed. What gets cached gets paid for once."

— Adapted

Chapter 10: Authentication, Billing, and Usage Limits

The API works. Your first call returned a response. Now you're moving toward something real — an internal tool, a product feature, a research pipeline that will make hundreds or thousands of calls. The question shifts from "can I make a call" to "how do I run this responsibly."

This chapter covers the operational side of the DeepSeek API: managing API keys at scale, the billing model in detail, working within rate limits, forecasting cost, and setting up the guardrails that prevent surprise bills.

API Key Management

You created a single API key in Chapter 4 to get started. For real use, you want more than one.

Why Multiple Keys

Any production deployment should use at least three keys, often more:

Per environment. Separate keys for `local-dev`, `staging`, and `production`. You can rotate or revoke keys for one environment without affecting the others, attribute spend to its source, and limit blast radius if a key leaks.

Per service or component. If multiple services use DeepSeek (the main app, a background worker, an evaluation harness), separate keys make billing legible and let you set per-service rate limits if your platform supports them.

Per developer. For dev teams, individual keys prevent the "who's burning through credit" mystery and let you revoke a developer's access cleanly when they leave.

The console makes creating and naming keys cheap. Use them generously.

Key Rotation

Plan to rotate keys periodically. Two reasons:

Compromise. Keys leak — committed to git accidentally, scraped from logs, exfiltrated by malware. Periodic rotation limits the exposure window if a leak goes undetected.

Personnel changes. When team members leave, their keys should be revoked. Per-developer keys make this trivial.

A reasonable cadence: quarterly rotation for production keys, plus immediate rotation on any suspicion of compromise. The mechanics are simple: create a new key, deploy it, validate everything works, then delete the old key.

Key Storage

Never hardcode keys. Never commit them to source control. Use:

Environment variables for local development, populated from a `.env` file kept out of version control via `.gitignore`.

Platform secrets management for production — AWS Secrets Manager, GCP Secret Manager, Azure Key Vault, HashiCorp Vault, your platform's equivalent.

CI/CD secrets for build pipelines that need to run evaluations or deployments — GitHub Actions secrets, GitLab CI variables, etc.

If a key ever appears in a screenshot, debugging output, a Slack message, or a public error report — revoke it immediately and rotate. There's no recovering an exposed key.

The Billing Model in Detail

Chapter 4 introduced the four-component pricing structure (input, output, cache hit, cache miss). Here's the operational depth.

What Counts as a Token

The model's tokenizer determines how text becomes tokens. For English, the rule of thumb is that one token is roughly three-quarters of a word. For other languages, code, structured data, and unusual text, the ratio differs:

- **Standard English prose**: ~0.75 word per token

- **Code (Python, JavaScript)**: often more tokens per character than prose, especially syntax-heavy languages
- **JSON**: token-inefficient because of all the brackets, commas, and quotes
- **Non-Latin languages**: usually more tokens per character than English
- **Whitespace, punctuation**: counts as tokens

For precise estimation before sending requests, use a local tokenizer library. The `tiktoken` library (Python, `pip install tiktoken`) handles this for OpenAI-compatible models including DeepSeek. JavaScript users have `gpt-tokenizer` and similar.

Input Pricing: Cache Hit vs. Cache Miss

Input pricing has two rates:

Cache miss — the standard input rate, charged when the input doesn't match a previously cached prefix.

Cache hit — a deeply discounted rate (often 80–90% off the cache-miss price), charged when the input begins with text that matches a previously cached prefix from a recent request.

The math matters. As of early 2026:

- V4 Flash cache miss: ~$0.14/M tokens
- V4 Flash cache hit: ~$0.0028/M tokens — that's a **50× discount**

For applications that share a system prompt or other prefix across many requests, hitting the cache repeatedly turns input cost into a rounding error. We cover the mechanics in detail in Chapter 12 — it earns its own chapter because it's that consequential.

Output Pricing

Output tokens cost more than input tokens. There's no cache hit/cache miss distinction for output (the model generates each output token from scratch).

For thinking-mode calls, the reasoning trace counts as output. A thinking-mode response of 3,000 reasoning tokens plus 500 final-answer tokens bills 3,500 output tokens. This is why thinking mode is a real cost decision, not just a UX one.

Tracking Cost Per Call

Every API response includes a `usage` field with token counts:

```
response.usage.prompt_tokens          # input tokens
response.usage.completion_tokens      # output tokens
response.usage.total_tokens           # sum
response.usage.prompt_tokens_details.cached_tokens  # how many input were cache
hits
```

For any production system, log these numbers per call. Aggregating over time tells you:

- Where your money is going (which endpoints, which use cases)
- Whether your cache hit rate is healthy
- Whether output tokens are creeping up (often a sign of prompt-engineering drift)
- Whether thinking mode is being used appropriately

A simple cost-per-call calculation:

```
def cost_cents(usage, model_pricing):
    cache_hits = usage.prompt_tokens_details.cached_tokens or 0
    cache_misses = usage.prompt_tokens - cache_hits
    output = usage.completion_tokens
    return (
        cache_hits * model_pricing["cache_hit_per_token"] +
        cache_misses * model_pricing["cache_miss_per_token"] +
        output * model_pricing["output_per_token"]
    ) * 100  # to cents
```

For larger deployments, build this attribution into your observability layer rather than ad-hoc.

Rate Limits

DeepSeek's API has rate limits at multiple levels:

Requests per minute (RPM). Distinct API calls per minute.

Tokens per minute (TPM). Total token throughput, summed across all calls.

Concurrent requests. In-flight requests at the same time.

The specific limits vary by account tier and may have changed since this was written. As of early 2026, free-tier accounts have modest limits suitable for development and small applications; paid tiers scale up as your usage history grows; enterprise customers get custom limits via direct contact.

Hitting a Rate Limit

When you exceed a limit, the API returns **429 Too Many Requests**. The standard pattern is to retry with exponential backoff:

```
import time
import random
from openai import RateLimitError

def call_with_retry(client, **kwargs):
    max_retries = 5
    for attempt in range(max_retries):
        try:
            return client.chat.completions.create(**kwargs)
        except RateLimitError:
            if attempt == max_retries - 1:
                raise
            sleep_seconds = (2 ** attempt) + random.random()
            time.sleep(sleep_seconds)
    raise RuntimeError("Exhausted retries")
```

The OpenAI client (which most users will use against DeepSeek's API) has retry helpers built in. Configure them rather than rolling your own.

Avoiding Rate Limits

For production workloads pushing against rate limits, a few strategies:

Batch where possible. If you're making N independent calls that could be combined into one (e.g., asking the model to process N items in a single response), one call uses one request slot instead of N.

Throttle on your side. A request queue with deliberate pacing keeps you from blasting the API and tripping the limit. Aim to stay below your TPM/RPM ceiling rather than hitting it.

Spread across keys cautiously. Each key has its own limits in some account configurations. Don't abuse this — it's not a way to multiply your effective rate limit, and DeepSeek may aggregate at the account level. But for distinct workloads, separate keys provide isolation.

Request a higher tier. For sustained high-volume production use, work with DeepSeek's commercial team to get appropriate limits.

Forecasting Cost

For new applications, cost forecasting is a recurring need. The basic approach:

1. **Estimate request volume per day.** How many calls will you make?
2. **Estimate token sizes** – average input and output tokens per call. Run a few sample calls and read the usage numbers.
3. **Estimate cache hit rate** – for prompts with shared structure, 70–90%; for one-off prompts, 0%.
4. **Apply pricing** – multiply token volumes by per-token rates and sum.

A worked example. Suppose you're building a customer support agent that uses DeepSeek-V4-Flash:

- 10,000 conversations per day
- Average 6 turns per conversation = 60,000 turns
- Average input tokens per turn = 1,500 (system prompt + history + user message)
- Average output tokens per turn = 300
- Estimated cache hit rate (system prompt is constant): ~70%

Daily token volume: - Input: 60,000 × 1,500 = 90M tokens (70% cache hits = 63M cache hits, 27M cache misses) - Output: 60,000 × 300 = 18M tokens

Daily cost (approximate): - Input cache hits: 63M × $0.0028/1M = $0.18 - Input cache misses: 27M × $0.14/1M = $3.78 - Output: 18M × $0.28/1M = $5.04 - **Total: ~$9 per day, ~$270 per month**

For comparison, the same workload on GPT-5.5 (without aggressive caching) would land somewhere around **$5,000–$10,000 per month**. The economics genuinely change.

Setting Up Spending Alerts and Hard Caps

Surprise bills are unpleasant. Three protective measures:

Spending alerts. The console supports email notifications at threshold spending levels (e.g., notify me at $50, $100, $500 in a month). Configure these. They're free; they prevent the worst surprises.

Hard caps where available. Some account configurations let you set a hard spending cap that pauses API access at the threshold. For dev/staging environments where a buggy script could otherwise burn through credit, this is invaluable. For production, a hard cap can cause outages — use carefully, with monitoring that alerts before the cap triggers.

Internal cost monitoring. Beyond DeepSeek's billing dashboard, implement your own tracking. A daily Slack message reporting yesterday's API cost is a cheap practice that catches issues fast.

Common Failure Modes and Their Costs

A few patterns that have burned real teams using DeepSeek and similar APIs:

Infinite retry loops on permanent errors. A bug that triggers a 4xx on every request, combined with naive retry logic, chews through tokens fast. Make sure your retry logic distinguishes transient (5xx, 429) from permanent (4xx) errors and only retries the former.

Runaway agent loops. An agent stuck in a thought loop can rack up substantial cost in minutes. Add hard step limits and total-cost circuits to anything agentic.

Production accidentally pointed at expensive model. Someone changes a config from `v4-flash` to `v4-pro` and forgets to revert. Costs jump 5–10× until someone notices. Per-environment monitoring and per-component tagging catch this.

Cache-invalidating prompt changes. A small edit to your system prompt (adding a date stamp, changing the date format) breaks every prefix cache. Cache hit rate drops from 90% to 0%; input costs grow 50×. Cache awareness when editing prompts is a habit worth developing — covered in Chapter 12.

Forgotten background jobs. A worker that's supposed to run nightly accidentally runs every minute. Costs accumulate quietly. Operational hygiene matters.

The Operational Discipline

Building on DeepSeek (or any commercial AI API) at scale demands operational discipline separate from prompt engineering or model selection:

- **Separate keys per environment and workload**
- **Periodic key rotation**
- **Logged usage metrics per call**
- **Cost attribution per use case**
- **Spending alerts and hard caps where appropriate**
- **Retry logic that distinguishes transient from permanent errors**
- **Awareness of cache behavior when editing prompts**
- **Step limits on agentic workloads**

None of this is glamorous. All of it is necessary for production deployments that don't blow up.

The next chapter returns to the fun part — choosing models in code, including the routing patterns that let a single application use V4 Pro for hard problems, V4 Flash for routine ones, and stay efficient across the mix.

Chapter 11: Choosing the Right Model in Code

In the web app, model selection is a UI choice. In the API, it's a string parameter. That sounds trivial — and for simple applications it is — but for any production workload at meaningful scale, model selection becomes one of the most consequential cost-and-quality levers you have.

This chapter covers the practical mechanics of choosing models in code: when to use each, how to switch dynamically, how to design routing patterns that pair the right model with each request, and how to handle the deprecations that come with using a fast-moving API.

The Decision: V4 Pro vs. V4 Flash

For most workloads, the relevant choice is between **DeepSeek-V4-Pro** and **DeepSeek-V4-Flash**. The model selector takes the model name as a string parameter:

```
response = client.chat.completions.create(
    model="deepseek-v4-pro",   # or "deepseek-v4-flash"
    messages=messages,
)
```

When V4 Pro wins:

- **Complex reasoning tasks** — multi-step logic, formal analysis, tasks where the answer depends on getting many things right at once
- **Production-quality writing** — anything you'd publish under your name without heavy editing
- **Code that must work** — non-trivial implementations where bugs are costly
- **Customer-facing responses** — where the quality difference is perceptible to end users
- **High-stakes decisions** — anything where a wrong answer has meaningful downstream consequences

When V4 Flash wins:

- **High-volume simple workloads** — classification, extraction, simple Q&A, summarization
- **Internal tooling** — where the engineering team can tolerate some quality variation
- **Evaluation harnesses** — running thousands of calls to test a system
- **Background processing** — non-time-critical work where cost matters more than marginal quality
- **Rapid prototyping** — iterating on prompts without burning through your budget

In most production deployments, the right answer isn't "always Pro" or "always Flash" — it's **routing** between them based on request shape. We cover routing patterns below.

Switching Models Mid-Conversation

Within a single conversation, you can switch the model on each call. The client doesn't care which model handled previous turns; it sends the conversation history to whichever model you specify on the current call.

A common pattern: **start with Flash for exploratory turns, escalate to Pro for the hard turn.**

```
def chat_with_escalation(conversation, message):
    conversation.append({"role": "user", "content": message})

    # Quick check: does this message look like it needs the strong model?
    if needs_pro(message):
        model = "deepseek-v4-pro"
    else:
        model = "deepseek-v4-flash"

    response = client.chat.completions.create(
        model=model,
        messages=conversation,
    )
    conversation.append(response.choices[0].message)
    return response, model
```

The `needs_pro()` heuristic can be as simple as keyword matching ("explain why," "design," "debug") or as sophisticated as a separate Flash call that classifies the request. For most applications, simple heuristics beat the complexity.

The Legacy Aliases

DeepSeek's API has supported two legacy model aliases:

- **`deepseek-chat`** — currently routes to V3.2
- **`deepseek-reasoner`** — currently routes to R1

These aliases are convenient for "I want the current best chat model" or "I want the current best reasoning model" — DeepSeek can update what they point to without breaking your code.

But they have downsides:

Behavior can change. When DeepSeek decides `deepseek-chat` should now point to V4-Flash instead of V3.2, your code's behavior shifts. Fine for most applications; a problem for anything that needs reproducibility.

Pricing can change. Different underlying models have different pricing. An alias swap can affect your bill.

Capabilities may shift. If V4 supports a feature V3.2 doesn't, behavior on the alias may newly succeed where it used to fail.

For new code, **use explicit model names** like `deepseek-v4-pro` or `deepseek-v4-flash`. The legacy aliases will eventually be deprecated; explicit names give you control over when you upgrade.

For production reproducibility (e.g., evaluation pipelines that need to be re-run next year and produce comparable results), pin to **specific version dates** if available — `deepseek-v4-pro-2026-04-24` or similar. This is the strongest guarantee against silent behavior changes.

Routing Patterns

For applications that handle a mix of request types, routing is where real cost optimization happens. A few patterns:

Heuristic Routing

The simplest pattern: classify each request based on keywords or simple features, route to Pro or Flash accordingly.

```
def route_model(message):
    text = message.lower()
    pro_signals = ["explain why", "design", "debug", "compare", "analyze", "plan
"]
    if any(signal in text for signal in pro_signals):
        return "deepseek-v4-pro"
    if len(text) > 500:  # long inputs might be analysis tasks
        return "deepseek-v4-pro"
    return "deepseek-v4-flash"
```

Heuristic routing is cheap to implement, fast (no extra API call), and easy to tune by inspecting routing decisions in production logs. For most applications, it's the right starting point.

Classifier Routing

For more nuanced routing, use a small Flash call to classify the request before routing the actual response:

```
def classify_and_route(message):
    classification = client.chat.completions.create(
        model="deepseek-v4-flash",
        messages=[
            {"role": "system", "content": "Classify this request as 'simple' or
'complex'. Respond only with the word 'simple' or 'complex'."},
            {"role": "user", "content": message},
        ],
        max_tokens=10,
    )
    label = classification.choices[0].message.content.strip().lower()
    return "deepseek-v4-pro" if "complex" in label else "deepseek-v4-flash"
```

This adds latency (one extra call) and a small cost, but captures nuance heuristics miss. Useful when classification quality matters.

Cascade Routing

Try the cheaper model first; escalate to the more expensive model only if the cheap model's answer fails some quality check.

```
def cascade(messages):
    # Try Flash first
    flash_response = client.chat.completions.create(
        model="deepseek-v4-flash",
        messages=messages,
    )
    answer = flash_response.choices[0].message.content

    if quality_check_passes(answer):
        return flash_response

    # Escalate to Pro
    return client.chat.completions.create(
        model="deepseek-v4-pro",
        messages=messages,
    )
```

The `quality_check_passes` function might check for specific failure patterns (the model said "I'm not sure"), look for required output structure, or use a separate Flash call to evaluate the answer. Cascade routing keeps cost low when Flash works and preserves quality when it doesn't.

Tier-Based Routing

For applications with explicit user tiers (free vs. paid, trial vs. enterprise), route by tier:

```
def route_by_tier(user_tier, message):
    if user_tier == "enterprise":
        return "deepseek-v4-pro"
    if user_tier == "paid":
        # Use cascade to escalate when needed
        return cascade_route(message)
    return "deepseek-v4-flash"
```

This is a business decision as much as an engineering one — but it's a common architecture for SaaS products built on AI.

Reasoning vs. Non-Reasoning Calls

Within V4, the orthogonal decision is whether to enable thinking mode. The same routing principles apply:

Heuristic. Enable thinking mode for messages matching certain patterns (math problems, multi-step questions, careful analysis).

Per-call. Enable it when you have specific reason to; leave it off by default.

Cascade. Try non-thinking first; escalate to thinking if the answer fails a quality check.

Combining model routing with thinking-mode routing gives four effective tiers:

1. Flash, no thinking (cheapest, fastest)
2. Flash, thinking (cheap, more careful)
3. Pro, no thinking (capable, fast)
4. Pro, thinking (most capable, slowest, most expensive)

For most applications, you'll use one or two tiers most of the time and the others as escalations. Designing the tiers explicitly into your routing logic makes the cost-quality decisions visible.

Handling Deprecations

DeepSeek has been generally good about deprecation notices, but eventually any model name you use today will be deprecated. A few practices:

Subscribe to the changelog. Watch DeepSeek's API documentation page or sign up for any update mailing list they offer. Deprecation announcements come with timelines.

Build deprecation-aware fallbacks. When a deprecated model returns an error, fall back to a current alternative rather than crashing.

Test new models proactively. When DeepSeek releases V5 (or V4.5, or whatever's next), try it against your eval set. Don't wait until your current model is deprecated to discover whether the upgrade works.

Centralize model selection. Keep model name strings in one place (a config file, a constants module). When you need to update across your codebase, you change one file rather than dozens.

```
# models.py
DEFAULT_FAST_MODEL = "deepseek-v4-flash"
DEFAULT_QUALITY_MODEL = "deepseek-v4-pro"
```

```
DEFAULT_REASONER = "deepseek-v4-pro"  # used to be deepseek-reasoner

# Anywhere else in code:
from models import DEFAULT_FAST_MODEL
client.chat.completions.create(model=DEFAULT_FAST_MODEL, ...)
```

A Worked Example: A Routed Chat Endpoint

Putting it together, a simple but production-realistic endpoint:

```
import os
from openai import OpenAI

client = OpenAI(
    api_key=os.environ["DEEPSEEK_API_KEY"],
    base_url="https://api.deepseek.com/v1",
)

DEFAULT_FAST_MODEL = "deepseek-v4-flash"
DEFAULT_QUALITY_MODEL = "deepseek-v4-pro"

PRO_SIGNALS = [
    "explain why", "explain how", "design", "debug",
    "compare", "analyze", "plan", "architect", "optimize",
]

def route_and_call(messages, force_thinking=False):
    last = messages[-1]["content"].lower()
    needs_pro = any(s in last for s in PRO_SIGNALS) or len(last) > 500

    model = DEFAULT_QUALITY_MODEL if needs_pro else DEFAULT_FAST_MODEL
    extra_body = {"thinking": {"type": "enabled"}} if force_thinking or needs_pr
o else {}

    response = client.chat.completions.create(
        model=model,
        messages=messages,
        extra_body=extra_body,
    )

    # Log routing for later analysis
    log_routing(model=model, thinking=bool(extra_body.get("thinking")),
                tokens=response.usage.total_tokens,
                cached=response.usage.prompt_tokens_details.cached_tokens or 0)
```

```
    return response

def log_routing(**fields):
    # Send to your observability layer
    ...
```

Roughly fifty lines that, over time, let you see exactly which requests went to which model, what the cost looked like, and where to refine the routing heuristics.

What This All Adds Up To

Model selection in code is the difference between an application that uses DeepSeek casually and one that uses it deliberately. The thoughtful applications:

- Use explicit model names, not legacy aliases
- Route between Flash and Pro based on request shape
- Layer thinking mode as an additional tier when reasoning matters
- Log routing decisions and costs per call
- Centralize model name strings for clean upgrades
- Test new models proactively before deprecations force the move

These practices add maybe a few hours of upfront engineering. They save weeks of confusion later when costs are unexpected, models are deprecated, or the team needs to understand why one part of the application is slow and expensive.

The next chapter is the highest-leverage cost optimization DeepSeek offers — and one of the most consequential things to understand about the API: prompt caching.

Chapter 12: Prompt Caching — The 80% Cost Saver

This is the most consequential chapter in Part III. If you skim everything else, read this carefully.

DeepSeek's API charges roughly 50× more for input tokens that miss the cache than for tokens that hit. A system that never benefits from caching pays one rate; a system designed to maximize hits pays something close to one-fiftieth as much for the same input volume. The difference compounds across every call, every day, for the lifetime of the application. Caching isn't a side optimization. For most production workloads, **it's the largest cost lever you have.**

This chapter covers what caching is, how it works under the hood, how to design prompts to maximize hits, the patterns that destroy hit rates, and a worked example showing the dollars-and-cents impact.

What Caching Is

DeepSeek's prompt caching lets the API server cache the model's processed state for a prefix of your input. When you send a subsequent request that begins with the same prefix, the server reuses the cached state rather than re-processing those tokens from scratch.

The benefit you see: those prefix tokens are billed at a deeply discounted rate (the **cache hit** rate), while the remaining tokens (anything new in your request) are billed at the standard rate (the **cache miss** rate).

A few key properties:

Caching is automatic. You don't enable it explicitly. The server handles it. You just pay differently depending on what you sent.

Caching is prefix-based. The cache matches from the start of your input. If your input begins with the same text as a recent request, those prefix tokens are cache hits. As soon as your input diverges, the rest is cache misses.

Cache lifetime is short but useful. Cached entries persist for a window (typically minutes to hours, varying by load). For a production system making frequent requests with shared prefixes, the cache stays warm.

Cache hits show up in usage data. The `usage.prompt_tokens_details.cached_tokens` field tells you how many input tokens were cache hits. Watching this field is how you know caching is working.

The Math

To put concrete numbers on it, using V4-Flash pricing as of early 2026:

- **Cache miss**: $0.14 per million input tokens
- **Cache hit**: $0.0028 per million input tokens

That's a **50× discount** for hits.

For a request with a 3,000-token prompt: - 100% cache miss: 3,000 × $0.14/1M = $0.00042 - 100% cache hit: 3,000 × $0.0028/1M = $0.0000084 - Difference: ~$0.0004 per call

Multiply by a million calls (a moderate production workload over a few months): the difference is ~$400 vs. ~$8. Across a year at higher volume: the difference is the price of an engineer.

For V4 Pro the per-token rates are higher but the ratio is similar. Percentage savings are about the same; absolute savings are larger.

How to Maximize Cache Hits

The discipline reduces to one principle: **put stable content first, variable content last.**

Cache hits depend on prefix matching. The server caches based on what comes at the start of your input. So:

Stable content (cacheable): - System prompts that don't change between calls - Few-shot examples that are part of your prompt template - Background documents or context that's reused - User-specific context that's the same across many calls (e.g., "this user is a Python developer with 5 years of experience...")

Variable content (not cacheable): - User messages that differ between calls - Per-call-specific data (timestamps, request IDs, dynamic values) - Conversation history (which grows with each turn)

The structure of an ideal cacheable prompt:

```
[ STABLE: System prompt (always the same) ]
[ STABLE: Few-shot examples (always the same) ]
[ STABLE: Per-user/per-session context that's stable for this user ]
[ VARIABLE: The current user message ]
```

Everything before the variable section is potentially cacheable. Everything from there onward is fresh.

Patterns That Work

A few specific patterns that produce high cache hit rates in practice:

The Long System Prompt + Short User Message

If you have a sophisticated system prompt (your application's "personality," instructions, output formatting requirements, examples) and the variable per-call content is just the user's question:

```
messages = [
    {"role": "system", "content": LONG_SYSTEM_PROMPT},  # cacheable
    {"role": "user", "content": user_message},          # variable
]
```

Cache hit rate on the system prompt approaches 100% across calls. Even at 2,000 tokens, you pay near-cache-hit rates for it from the second call onward.

Stable RAG Context First

For Retrieval-Augmented Generation (RAG) workflows where you fetch documents and prepend them to the prompt:

```
messages = [
    {"role": "system", "content": SYSTEM_PROMPT},
    {"role": "system", "content": f"Reference document:\n{retrieved_doc}"},  # c
acheable per-doc
    {"role": "user", "content": user_question},
]
```

For users who query the same retrieved document multiple times in a session, the document context becomes a cache hit from the second call onward.

Stable Conversation Prefix in Chat

For chatbots, the conversation history grows with each turn. The early turns become a stable prefix relative to the latest turn:

```
# Turn 1: messages = [system, user_1]
# Turn 2: messages = [system, user_1, assistant_1, user_2]
# Turn 3: messages = [system, user_1, assistant_1, user_2, assistant_2, user_3]
```

For Turn 3, the prefix `[system, user_1, assistant_1, user_2, assistant_2]` was sent on Turn 2's request and may still be cached. The new tokens (the user_3 message) are cache misses; everything before is cache hits.

This is why long-running chat conversations actually get cheaper per turn over time: the prefix grows but is increasingly cached.

Patterns That Destroy Hit Rates

Equally important: patterns that look fine but quietly bust the cache.

Variable Content at the Start

A timestamp, request ID, or session identifier at the start of your prompt invalidates the entire cache:

```
# BAD — every request has a different timestamp at the start
system = f"Current time: {datetime.now()}\n\nYou are a helpful assistant..."
```

Even if the rest of the system prompt is identical across calls, the changing timestamp means the prefix never matches. Every call is 100% cache miss.

The fix: move variable content to the user message, or to a later position in the prompt.

Conversation History Reordering

If you reorder the conversation history between turns (e.g., showing the most recent first instead of chronologically), each call has a different prefix:

```
# Turn 1: [system, user_1]
# Turn 2: [system, user_2, user_1]  # reordered — cache busted
```

Don't reorder. Append.

Per-User Variation in System Prompt

If you have a system prompt that includes per-user information at the very start:

```
# BAD
system = f"You are talking to user_id={user_id}, name={user.name}.\n\n[generic i
nstructions]"
```

Cache hits work per-user (each unique prefix gets its own cache entry), but with many users and few requests per user, the cache barely warms up before the entry expires.

The fix: generic instructions first, per-user variation later:

```
# BETTER
system = f"[generic instructions]\n\nUser context: id={user_id}, name={user.name}
"
```

Now the generic part hits a single cache entry across all users.

Editing Prompts Casually

Production prompt iteration is a constant temptation: "let me just tweak the system prompt." Every tweak invalidates the cache. Yesterday's built-up cache is gone.

This isn't an argument against improving prompts — it's an argument for being deliberate. Plan prompt changes; deploy them at low-traffic times if possible; expect a cache-miss spike for the first hour or so afterward.

Including Random Noise

Some applications include random tokens (UUIDs, randomly shuffled examples, jitter) in their prompts thinking it'll help with diversity. It doesn't help — and it destroys caching.

If you need diversity, get it from temperature/sampling parameters, not from prompt content variation.

Verifying Caching Is Working

The `usage.prompt_tokens_details.cached_tokens` field is your friend. After every call:

```
response = client.chat.completions.create(...)
total_input = response.usage.prompt_tokens
cached = response.usage.prompt_tokens_details.cached_tokens or 0
cache_hit_rate = cached / total_input if total_input else 0
print(f"Cache hit rate: {cache_hit_rate:.1%}")
```

For a healthy production workload, expect:

- **80%+ cache hit rate** for chatbot conversations after the first few turns
- **50%+ cache hit rate** for RAG workloads with stable system prompts
- **0% cache hit rate** for one-off, never-repeating queries (this is expected; not a bug)

If you expect a high hit rate and see a low one, the most likely cause is variable content sneaking into the prefix. Audit prompt construction.

A Worked Example: 90% Cost Reduction on a Real Workload

A concrete example, sized for a small-to-medium deployment:

You're running a customer support chatbot. 100,000 conversations per day. Four turns per conversation on average. Average input per turn: 2,000 tokens. Average output per turn: 200 tokens.

Without caching awareness:

Daily volume: - Input: 100,000 × 4 × 2,000 = 800M tokens, all cache miss - Output: 100,000 × 4 × 200 = 80M tokens

Daily cost (V4 Flash): - Input: 800M × $0.14/1M = **$112** - Output: 80M × $0.28/1M = $22.40 - **Total: ~$134/day**

With caching awareness (system prompt of 1,500 tokens stable, only the user message and growing history variable):

For each conversation's four turns, roughly: - Turn 1: 100% miss on the 1,500 system prompt + 500-token user message - Turn 2: cache hit on system prompt (1,500), miss on assistant_1 (200) + user_2 (500) - Turn 3: cache hit on system + user_1 + assistant_1, miss on user_2 + assistant_2 + user_3 - ...

Across four turns averaging 2,000 input tokens each, roughly 70% becomes cache hit (the stable system prompt and the accumulated history).

Daily volume: - Input cache hits: 800M × 0.70 = 560M tokens - Input cache misses: 800M × 0.30 = 240M tokens - Output: 80M tokens

Daily cost: - Cache hits: 560M × $0.0028/1M = $1.57 - Cache misses: 240M × $0.14/1M = $33.60 - Output: 80M × $0.28/1M = $22.40 - **Total: ~$58/day**

Savings: $76/day, $2,280/month, $27,000/year. Same workload, same model, just by structuring the prompt to maximize cache hits.

For larger deployments, the savings scale directly. A team running 10× this volume saves $270K/year. A team running 100× saves $2.7M.

This is why caching earns its own chapter.

When Caching Doesn't Apply

Some workloads genuinely can't benefit:

- **One-off queries** where each request is completely unrelated
- **Workloads with naturally high prompt variance** — every request structurally different
- **Workloads with deliberate prompt randomization** — for some research applications, you want the diversity

For these, the caching savings aren't available. The model and pricing are still cheap relative to alternatives, but the order-of-magnitude advantage requires the cacheable structure.

For most production applications — chatbots, agents, RAG systems, content generation pipelines — there's substantial room for caching. The discipline pays for itself many times over.

What to Do Next

If you're building on the DeepSeek API and haven't designed for caching, an audit-and-refactor pass is one of the highest-ROI things you can do:

1. **Measure your current cache hit rate** by logging the `cached_tokens` field across recent calls
2. **Identify variable content in your prefix** that could be moved later
3. **Refactor prompt construction** to put stable content first
4. **Re-measure** after deploying to see the improvement
5. **Update cost forecasts** based on the new hit rate

Rarely glamorous. Almost always worth doing.

The next chapter covers another foundational API capability: tool calling, where DeepSeek can request that your application execute functions and return their results, enabling agent-style workflows.

Chapter 13: Tool Calling and Function Use

Language models on their own can't fetch live data, query your database, send an email, run code, or take any action in the world outside their text generation. They write text. That's it.

Tool calling is the protocol that bridges the gap. You define a set of tools (functions) the model can request to use; when the model judges a tool would help, it returns a structured request to call one; your application executes the call and returns the result; the model continues its response, now informed by the tool output.

This chapter covers how tool calling works on the DeepSeek API, when to use it, single-tool and multi-tool patterns, and the agent-loop pattern that's become standard for LLM-driven workflows.

How Tool Calling Works

The basic dance has four steps:

1. **You define tools.** A list of available functions, each with a name, description, and JSON schema for its arguments.
2. **You make a chat completion request** with the user's message and the tool definitions.
3. **The model responds** with either a normal text answer (if no tool is needed) or a structured tool call request (if a tool would help).
4. **You execute the tool** and feed the result back to the model in a follow-up request. The model uses the result to produce its final answer.

The key insight: the model doesn't call any tools itself. It returns a structured request to call a tool, your code executes it, and you feed the result back. This keeps the model in its lane (text generation) while letting your application do the world-affecting work safely.

A Minimal Example

A canonical example: weather lookup.

```
import json
from openai import OpenAI

client = OpenAI(
    api_key=os.environ["DEEPSEEK_API_KEY"],
    base_url="https://api.deepseek.com/v1",
)

# Define the tool schema
tools = [
    {
        "type": "function",
        "function": {
            "name": "get_weather",
            "description": "Get the current weather for a city.",
            "parameters": {
                "type": "object",
                "properties": {
                    "city": {
                        "type": "string",
                        "description": "City name, e.g. 'San Francisco'",
                    },
                    "unit": {
                        "type": "string",
                        "enum": ["celsius", "fahrenheit"],
                        "default": "fahrenheit",
                    },
                },
                "required": ["city"],
            },
        },
    },
]

# The actual tool implementation
def get_weather(city, unit="fahrenheit"):
    # In real life this hits a weather API; mocked here:
    return {"city": city, "temp": 72, "unit": unit, "conditions": "sunny"}

# First call: send the user message + tool definitions
```

```
messages = [{"role": "user", "content": "What's the weather in Boston?"}]
response = client.chat.completions.create(
    model="deepseek-v4-flash",
    messages=messages,
    tools=tools,
)

assistant_message = response.choices[0].message
messages.append(assistant_message)

# Did the model want to call a tool?
if assistant_message.tool_calls:
    for tool_call in assistant_message.tool_calls:
        if tool_call.function.name == "get_weather":
            args = json.loads(tool_call.function.arguments)
            result = get_weather(**args)
            messages.append({
                "role": "tool",
                "tool_call_id": tool_call.id,
                "content": json.dumps(result),
            })

    # Second call: send the conversation again, now with tool results
    final_response = client.chat.completions.create(
        model="deepseek-v4-flash",
        messages=messages,
    )
    print(final_response.choices[0].message.content)
else:
    # The model answered directly without tools
    print(assistant_message.content)
```

The model recognizes that "What's the weather in Boston?" benefits from `get_weather`, returns a structured request with `city: "Boston"`, your code executes the lookup, you return the result, and the model produces a natural-language response.

When to Use Tools

Tool calling is appropriate when:

The model needs information it doesn't have. Current data, database lookups, document retrieval, web search — anything where the answer depends on something outside training data.

The model needs to take action. Sending an email, creating a calendar event, posting a comment, updating a record — anything that affects state in the world.

The model needs to use a deterministic function. Math computations where you want exact answers, code execution, regex matching — anything where you'd rather call real code than trust the model to simulate it.

The model benefits from structured intermediate steps. Multi-step workflows where each step's output informs the next.

Tool calling is overkill when:

- The model can answer directly from training data
- The "tool" is just a fancy way to format an answer the model already has
- You have only one possible tool and would always call it (just call it without involving the model in the decision)

Defining Tools Well

The quality of your tool definitions strongly affects the model's ability to use them correctly. A few practices:

Clear, specific descriptions. "Get the current weather for a city" beats "Weather function." The description tells the model when this tool is appropriate.

Type-strict parameters. Use the JSON schema to constrain what the model can pass. Enums for fixed-value parameters, integer types for numeric parameters, required vs. optional flags.

Descriptions on parameters. Each parameter gets its own. The model uses these to fill them correctly.

Conservative tool count. Don't define dozens of tools for the model to choose between. The more tools, the more likely the model picks the wrong one or fails to pick any. Define what's needed; consolidate where possible.

Predictable error handling. When a tool fails, return the error in a structured form the model can react to. "Tool failed" is less useful than `{"error": "city_not_found", "suggestion": "Did you mean Boston?"}`.

Parallel Tool Calls

DeepSeek-V4 supports parallel tool calling — the model can request multiple tool calls in a single response when they're independent.

```
# User message: "What's the weather in Boston, New York, and Chicago?"
# The model returns three parallel tool_call objects, one per city.
```

Your code can execute these in parallel:

```
import asyncio

async def execute_tool_call(tool_call):
    if tool_call.function.name == "get_weather":
        args = json.loads(tool_call.function.arguments)
        # ... call weather API asynchronously
        return result

async def execute_all(assistant_message):
    results = await asyncio.gather(*[
        execute_tool_call(tc) for tc in assistant_message.tool_calls
    ])
    return [
        {"role": "tool", "tool_call_id": tc.id, "content": json.dumps(r)}
        for tc, r in zip(assistant_message.tool_calls, results)
    ]
```

For workflows with multiple independent lookups, parallel execution is much faster than serial.

The Agent Loop

For workloads that need multiple rounds of tool use — call a tool, see the result, call another tool, see that result, and so on — the standard pattern is the **agent loop**:

```
def run_agent(user_message, max_steps=10):
    messages = [{"role": "user", "content": user_message}]
```

```python
    for step in range(max_steps):
        response = client.chat.completions.create(
            model="deepseek-v4-flash",
            messages=messages,
            tools=tools,
        )
        assistant_message = response.choices[0].message
        messages.append(assistant_message)

        if not assistant_message.tool_calls:
            # No more tool calls — the model is done
            return assistant_message.content

        # Execute every tool call
        for tool_call in assistant_message.tool_calls:
            result = execute_tool(tool_call)
            messages.append({
                "role": "tool",
                "tool_call_id": tool_call.id,
                "content": json.dumps(result),
            })

    raise RuntimeError("Agent didn't terminate within max_steps")
```

A few important details:

`max_steps` is non-negotiable. Without a step limit, a misbehaving agent loop runs indefinitely, racking up cost and time. Pick a reasonable limit for your workload (typically 5–20) and enforce it.

Cost circuit breakers help too. Beyond step count, monitor per-conversation token spend and abort if it exceeds a budget.

The model decides when to stop. When the assistant message has no tool calls, that's the model's signal that it has its final answer.

Errors should be feedback to the model. When a tool fails, return the error to the model rather than crashing. The model often recovers gracefully — "the city wasn't found, let me try a different spelling."

Combining Tools with Thinking Mode

Tool calling and thinking mode are orthogonal. You can enable both, neither, or either alone. For agent workloads, combining them often produces better results:

```
response = client.chat.completions.create(
    model="deepseek-v4-pro",
    messages=messages,
    tools=tools,
    extra_body={"thinking": {"type": "enabled"}},
)
```

The model produces explicit reasoning about which tool to use and why, then executes the call. For complex multi-tool workflows, this dramatically improves correctness.

The cost is higher (thinking mode generates more output tokens), but for production agents, the quality gains usually justify it.

Common Tool-Calling Patterns

A few patterns that show up in real applications:

Database Query Tool

```
tools = [{
    "type": "function",
    "function": {
        "name": "query_database",
        "description": "Run a read-only SQL query against the user database.",
        "parameters": {
            "type": "object",
            "properties": {
                "query": {"type": "string", "description": "A SQL SELECT stateme
nt."},
            },
            "required": ["query"],
        },
    },
}]
```

In your tool implementation: parse the SQL, validate it's a SELECT, execute against a read-only connection, return results. **Never** let the model write arbitrary SQL against a writable database — read-only contexts only, with careful validation.

RAG Search Tool

```
tools = [{
    "type": "function",
    "function": {
        "name": "search_knowledge_base",
        "description": "Search the company knowledge base for relevant documents.
",
        "parameters": {
            "type": "object",
            "properties": {
                "query": {"type": "string", "description": "Natural language sea
rch query."},
                "top_k": {"type": "integer", "default": 5},
            },
            "required": ["query"],
        },
    },
}]
```

For applications where the model needs to look up information from a corpus, a search tool is cleaner than stuffing everything into the context.

Action Tool with Confirmation

```
tools = [{
    "type": "function",
    "function": {
        "name": "send_email",
        "description": "Compose and send an email. Requires user confirmation be
fore sending.",
        "parameters": {
            "type": "object",
            "properties": {
                "to": {"type": "string"},
                "subject": {"type": "string"},
                "body": {"type": "string"},
                "confirmed": {"type": "boolean", "description": "Set to true onl
y after explicit user confirmation."},
            },
            "required": ["to", "subject", "body", "confirmed"],
```

```
        },
    },
}]
```

For action tools (anything that affects the world), build confirmation into the schema. The model then needs to ask the user before sending, and the tool implementation can hard-check the `confirmed` field.

Failure Modes to Watch For

Tool calling is powerful but can fail in characteristic ways:

Wrong tool selection. The model picks a tool that doesn't fit. Mitigations: clearer descriptions, fewer tools, thinking mode enabled.

Wrong arguments. The model calls the right tool with bad arguments – mistyped parameters, missed required fields. Mitigations: strict JSON schema, defensive validation in your tool implementation.

Hallucinated tool results. Less common with proper tool calling (the model is supposed to wait for actual results), but possible if you're not handling the protocol cleanly. Make sure tool result messages have the correct `tool_call_id` and the loop strictly waits for results before continuing.

Infinite tool loops. The model keeps calling tools without ever producing a final answer. Mitigations: hard `max_steps` limit, cost circuits, prompt design that encourages the model to commit to an answer.

Errors interpreted as content. If your tool returns errors as plain strings, the model may interpret them as content rather than failure signals. Use structured error objects (`{"error": "..."}`) so the model recognizes them as errors.

What Tool Calling Enables

Tool calling is the foundational capability that turns LLMs from text-generators into systems that do things:

- **Chatbots that can take actions** (book appointments, send messages, update records)
- **Research agents** that can search, fetch, and synthesize across sources

- **Coding agents** that can read files, run tests, and propose changes
- **Customer support assistants** that can look up account info and execute requests
- **Workflow automation** where natural-language requests trigger structured actions

The patterns in this chapter are the building blocks for all of these. We'll return to agent workflows specifically in Chapter 18.

The next chapter covers another structured-output capability often paired with tool calling: JSON mode, which lets you reliably get machine-parseable responses from the model.

Chapter 14: JSON Mode and Structured Outputs

For applications that consume model output programmatically — extracting fields from documents, classifying inputs, populating database records — free-form text is the wrong shape. You need structured data.

This chapter covers JSON mode and the broader structured-output patterns: how to reliably get machine-parseable responses from DeepSeek, how to enforce schemas, and how to handle the cases where the model doesn't quite cooperate.

What JSON Mode Does

JSON mode is a parameter on a chat completion request that tells the model: "Your response must be valid JSON." When enabled, the model is constrained at decoding time to produce output that parses cleanly.

```
response = client.chat.completions.create(
    model="deepseek-v4-flash",
    messages=[
        {"role": "system", "content": "Extract name, email, and phone from the u
ser's input. Respond as JSON."},
        {"role": "user", "content": "Hi, I'm Jane Doe. You can reach me at jane@
example.com or 555-1234."},
    ],
    response_format={"type": "json_object"},
)

import json
data = json.loads(response.choices[0].message.content)
# data == {"name": "Jane Doe", "email": "jane@example.com", "phone": "555-1234"}
```

A few important details:

You still need to ask for JSON. Setting `response_format` doesn't tell the model what JSON to produce. Your prompt has to specify the structure. JSON mode just guarantees the response will parse.

The output is always parseable. You won't see truncated responses, missing commas, or other syntactic errors that plague text-based JSON extraction.

The output isn't always semantically correct. Valid JSON can still have wrong values, missing fields, or unexpected schema. JSON mode prevents parse errors, not validation errors.

Why You Want Structured Output

Without structured output, you're parsing free-form text and hoping the model wrote it the way you expected. With structured output, you have a contract.

The benefits compound:

- **No regex extraction.** No fragile string-matching to pull out the values you need.
- **No retry loops to clean up text formatting.** Ask for a list, get prose, and now you're writing a parser.
- **No silent failures.** Invalid JSON throws on parse — visible — instead of "we extracted the wrong thing and didn't notice."
- **Easier testing.** Structured output is easier to assert against than free-form text.
- **Easier composition.** The output of one structured call feeds cleanly into the next call or into application logic.

For any application that processes model output programmatically, structured output is the right default.

JSON Mode Patterns

A few patterns that work in practice:

Field Extraction

The most common use: pull specific fields from unstructured input.

```
SYSTEM_PROMPT = """Extract the following fields from the user's input. Respond a
s JSON.

Required fields:
```

```
- title: The job title mentioned (string, or null if not present)
- company: The company name (string, or null if not present)
- start_date: When the role started (ISO 8601 date string, or null)
- end_date: When the role ended (ISO 8601 date string, or null if current)

Return only the JSON object, nothing else.
"""

response = client.chat.completions.create(
    model="deepseek-v4-flash",
    messages=[
        {"role": "system", "content": SYSTEM_PROMPT},
        {"role": "user", "content": resume_excerpt},
    ],
    response_format={"type": "json_object"},
)
fields = json.loads(response.choices[0].message.content)
```

Classification

Returning a category (or set of categories) from a fixed taxonomy:

```
SYSTEM_PROMPT = """Classify the support ticket into one of: billing, technical,
account, feature_request, other.

Respond as JSON: {"category": "<one of the above>", "confidence": <0.0 to 1.0>,
"rationale": "<brief explanation>"}
"""

response = client.chat.completions.create(
    model="deepseek-v4-flash",
    messages=[
        {"role": "system", "content": SYSTEM_PROMPT},
        {"role": "user", "content": ticket_body},
    ],
    response_format={"type": "json_object"},
)
result = json.loads(response.choices[0].message.content)
# result["category"], result["confidence"], result["rationale"]
```

The confidence score and rationale serve two purposes: downstream routing (low-confidence cases go to human review) and evaluating the model's behavior over time.

Multi-Item Extraction

Returning a list of items from a document:

```
SYSTEM_PROMPT = """Extract every meeting mentioned in the email. Respond as JSON:
 {"meetings": [{"title": "...", "datetime": "...", "attendees": [...]}, ...]}

If no meetings are mentioned, return {"meetings": []}.
"""
```

Lists are where prose-based parsing tends to break down most. JSON mode handles it cleanly.

Hierarchical Output

For nested structures (analyses, plans, reports with sections):

```
SYSTEM_PROMPT = """Analyze the document and respond as JSON:
{
  "summary": "<one-paragraph summary>",
  "key_points": ["<point>", "<point>", ...],
  "concerns": [
    {"concern": "<description>", "severity": "<low|medium|high>", "evidence": "<
quote>"}
  ],
  "recommendations": ["<recommendation>", ...]
}
"""
```

The more complex your schema, the more important it becomes to specify it clearly in the prompt.

Schema Specification: Loose vs. Strict

JSON mode itself just guarantees parseability. To make the model follow a specific schema (specific field names, types, constraints), you have a few options:

Inline Schema Description (Loose)

The simplest approach: describe the schema in the system prompt as shown above. Works for most cases but isn't bulletproof — the model occasionally drifts.

JSON Schema in the Prompt (Tighter)

Include an actual JSON Schema definition in your prompt:

```
SYSTEM_PROMPT = f"""Extract fields from the input. Respond as JSON conforming to
 this schema:

{json.dumps(my_schema, indent=2)}

Required fields must be present. Use null for optional fields if data is missing.
"""
```

The model reads the schema and tries to honor it. Better than loose description, but still not strictly enforced.

Schema-Constrained Generation (Strict, when supported)

For some models and providers, you can pass a schema directly to the API and get strict enforcement at decoding time:

```
response = client.chat.completions.create(
    model="deepseek-v4-pro",
    messages=messages,
    response_format={
        "type": "json_schema",
        "json_schema": {
            "name": "extracted_fields",
            "schema": my_schema,
            "strict": True,
        },
    },
)
```

When supported, this is the strongest guarantee — output is mathematically constrained to match the schema. Check current DeepSeek documentation for support and exact syntax; schema-constrained generation has been evolving across providers.

Validation as the Belt-and-Suspenders Layer

Even with strict schema enforcement, validate output server-side. Parse it, check required fields, validate types and constraints, fall back gracefully on validation failure.

```python
from pydantic import BaseModel, ValidationError

class TicketClassification(BaseModel):
    category: str
    confidence: float
    rationale: str

try:
    raw = json.loads(response.choices[0].message.content)
    classification = TicketClassification(**raw)
except (json.JSONDecodeError, ValidationError) as e:
    # Fall back: log, retry with stricter prompt, route for human review, etc.
    handle_validation_failure(e, response)
```

`pydantic` (Python) and similar libraries in other languages give you declarative schema validation. Use them. The alternative is hand-written validation logic that always seems to miss the case that comes up in production.

Patterns That Don't Work Well

A few things JSON mode and structured output don't do well:

Free-form prose plus structured fields. If you want a long natural-language answer AND structured fields, split it into two calls — one for the prose, one for the fields. Asking for both in a single response usually produces awkward results.

Very large JSON outputs. When the response would be many thousands of tokens of structured data, the model sometimes truncates or stops mid-structure. For large extractions, batch into smaller calls.

Schemas with complex constraints the model can't reason about. "Each item's price field must be greater than the previous item's" is hard to enforce through schema alone. For constraints like this, validate after the fact and ask the model to fix.

Outputs requiring math. "Calculate the average and put it in the result." The model will do its best but may produce wrong arithmetic. For exact computation, use a tool call (Chapter 13) or compute it in your code after extraction.

Combining JSON Mode with Caching

JSON mode pairs well with prompt caching. The fixed schema description in your system prompt is highly cacheable — set it once, get cache hits across all calls. The only fresh portion is the per-call user input.

For high-volume extraction workloads, JSON mode plus cache-friendly prompt design delivers both reliable structured output AND minimal cost.

A Worked Example: Document Triage Pipeline

Putting it together — a small document triage application that processes incoming documents and produces a structured triage decision:

```
import os
import json
from openai import OpenAI
from pydantic import BaseModel, ValidationError
from typing import Literal

client = OpenAI(
    api_key=os.environ["DEEPSEEK_API_KEY"],
    base_url="https://api.deepseek.com/v1",
)

class TriageDecision(BaseModel):
    document_type: Literal["contract", "invoice", "report", "letter", "other"]
    urgency: Literal["low", "medium", "high"]
    summary: str
    suggested_action: str
    confidence: float

SYSTEM_PROMPT = """You are a document triage assistant.

Analyze the document and respond as JSON:
{
  "document_type": "contract" | "invoice" | "report" | "letter" | "other",
  "urgency": "low" | "medium" | "high",
  "summary": "<two-sentence summary>",
  "suggested_action": "<what should happen next, in one sentence>",
  "confidence": <0.0 to 1.0, your confidence in the classification>
}
```

```
Return only the JSON, nothing else."""

def triage(document_text):
    response = client.chat.completions.create(
        model="deepseek-v4-flash",
        messages=[
            {"role": "system", "content": SYSTEM_PROMPT},
            {"role": "user", "content": document_text},
        ],
        response_format={"type": "json_object"},
    )
    raw = json.loads(response.choices[0].message.content)
    try:
        decision = TriageDecision(**raw)
    except ValidationError as e:
        # Log, route for human review
        return None
    return decision

# Usage
decision = triage(some_document)
if decision and decision.urgency == "high":
    notify_team(decision)
```

A complete production-realistic pipeline in about 40 lines of code. Cache hits on the system prompt drop per-document cost to near-zero. Pydantic validation catches schema drift. The output integrates cleanly with downstream logic.

What Structured Output Enables

For applications that need to extract, classify, route, transform, or otherwise process information at scale, structured output is the foundation:

- **Document processing pipelines** (invoices, contracts, resumes, reports)
- **Data extraction** from web pages, emails, tickets, transcripts
- **Classification systems** that route inputs to different handlers
- **Scoring and rating** with confidence values
- **Workflow automation** where AI decisions trigger structured downstream actions
- **Data labeling** for ML training pipelines

Almost every production AI application has at least one structured-output node in its workflow. The patterns in this chapter are the building blocks.

The next chapter covers a different kind of structured output — Fill-In-the-Middle (FIM) completion, used for code completion and document infilling.

Chapter 15: FIM Completion (Fill-In-the-Middle)

Most chat completion use cases follow the same pattern: you provide a prompt that ends with the user's question, and the model generates tokens that follow. The model writes after your input.

Fill-In-the-Middle (FIM) completion flips that pattern. You provide a prefix AND a suffix, and the model fills in the content between them. The model writes in the middle.

It's a small-feeling feature with outsized importance for specific workloads — particularly code completion and document repair. This chapter covers what FIM is, when it's the right tool, the API mechanics, and the patterns that make it useful.

What FIM Is

In FIM mode, instead of "here's text, write what comes next," you tell the model: "here's text before the cursor, here's text after the cursor, fill in what goes between."

A simple example with code:

```
prefix:  def calculate_total(items):
            total = 0
            for item in items:

suffix:      return total

# Model fills in: total += item.price
```

The model sees both sides of the gap and produces content that bridges them — code that fits the surrounding context, not just a continuation of the prefix.

When FIM Is the Right Tool

FIM shines when surrounding context should constrain what gets generated:

Code completion. A user is editing a file. The cursor sits mid-function. The lines before and after should inform the suggestion. This is the classic IDE autocomplete scenario.

Document repair. A document has a missing or corrupted section. You know what comes before and after; you need to reconstruct the middle.

Template filling. A template has placeholders that need context-appropriate content. The surrounding template constrains what's reasonable.

Code refactoring with context. You're rewriting one function in a file. The other functions around it are context that should inform the rewrite.

FIM is the wrong tool when:

- You only have a prefix (use standard chat completion)
- The middle content should be free-form and unconstrained by what comes after
- You're doing creative generation where surrounding context shouldn't be limiting
- The "middle" is actually most of the response (the prefix and suffix are just context, and standard chat completion is more natural)

The API: Non-Thinking Mode Only

DeepSeek's FIM endpoint requires **non-thinking mode**. Thinking mode and FIM aren't compatible — FIM is built for fast, context-aware completion, not for explicit chain-of-thought reasoning.

In practice this is a feature, not a limitation. FIM use cases (code completion, document infilling) reward speed and don't typically benefit from reasoning traces. The constraint matches the use case.

The Request Shape

DeepSeek's FIM endpoint uses a specific completions API (separate from the chat completions API) with `prompt` and `suffix` fields:

```
import os
from openai import OpenAI
```

```
client = OpenAI(
    api_key=os.environ["DEEPSEEK_API_KEY"],
    base_url="https://api.deepseek.com/beta",  # FIM may live at a beta endpoint
)

response = client.completions.create(
    model="deepseek-v4-flash",
    prompt="def calculate_total(items):\n    total = 0\n    for item in items:\n
        ",
    suffix="    return total\n",
    max_tokens=50,
)

print(response.choices[0].text)
# Likely produces: total += item.price
```

A few mechanical notes:

The endpoint is the legacy `/completions`, not `/chat/completions`. The OpenAI client's `client.completions.create` method targets the right endpoint.

`prompt` is what comes before the cursor; `suffix` is what comes after. The model generates content that fits between them.

`max_tokens` limits generated content length. For autocomplete-style use, keep this small (20–100 tokens) to prevent runaway generation.

No `messages` array. FIM is a completions API, not a chat API. No conversation history — just prefix + suffix.

Check current docs for endpoint paths. DeepSeek's FIM endpoint location and exact API shape have evolved. The basics above hold; specifics may have shifted.

Use Cases in Detail

Code Autocomplete

The canonical FIM use case. An editor (or CLI tool, or any code-editing context) has a cursor position. Everything before and after the cursor is context.

```
def code_complete(file_text, cursor_position, max_tokens=80):
    prefix = file_text[:cursor_position]
    suffix = file_text[cursor_position:]

    response = client.completions.create(
        model="deepseek-v4-flash",
        prompt=prefix,
        suffix=suffix,
        max_tokens=max_tokens,
        stop=["\n\n", "\n}"],  # stop at logical breakpoints
    )
    return response.choices[0].text
```

For low-latency editor integrations, V4 Flash is the right model. Lower latency and lower cost matter more than the marginal quality difference for suggestions the user will accept or reject within a second.

Document Infilling

A document has a known prefix and suffix; you need to generate the middle:

```
def infill_section(document_text, before_marker, after_marker, max_tokens=500):
    before_idx = document_text.find(before_marker)
    after_idx = document_text.find(after_marker, before_idx)

    prefix = document_text[:before_idx + len(before_marker)] + "\n"
    suffix = "\n" + document_text[after_idx:]

    response = client.completions.create(
        model="deepseek-v4-flash",
        prompt=prefix,
        suffix=suffix,
        max_tokens=max_tokens,
    )
    return response.choices[0].text
```

Useful for templated documents (legal contracts, structured reports) with stable structure but variable content.

Refactoring with Surrounding Context

When refactoring one function in a larger file, FIM gives the model the surrounding code as context:

```
def refactor_function(file_text, function_start, function_end, instructions):
    prefix = file_text[:function_start] + "\n# Refactor: " + instructions + "\n"
    suffix = file_text[function_end:]

    response = client.completions.create(
        model="deepseek-v4-pro",  # Pro for refactor-quality
        prompt=prefix,
        suffix=suffix,
        max_tokens=1000,
    )
    return response.choices[0].text
```

The model sees what the surrounding code is doing — naming conventions, style, dependencies — and produces a refactor that fits.

Comparing FIM to Standard Chat Completion

For code-related work, you can often choose between FIM and standard chat completion. When does each win?

Standard chat completion wins when:

- You want to give the model explicit instructions ("refactor this for performance")
- The conversation is multi-turn (user asks, model responds, user follows up)
- You want explanations alongside code
- You're using thinking mode for careful reasoning
- The output should be free-form, not constrained by a "suffix"

FIM completion wins when:

- You're doing autocomplete-style suggestion (low latency, immediately consumable)
- The surrounding context strongly constrains what's appropriate
- You don't want explanations, just the filled-in content
- You're scaling to high request volume where chat completion overhead matters

For most production code-completion workflows (IDE plugins, real-time autocomplete in CLIs), FIM is the right primitive. For code conversations and explained refactors, chat completion is.

Cost Considerations

FIM completion is priced the same as standard completion — per-token input and output rates apply.

For high-volume autocomplete workloads, two cost optimizations matter:

Keep `max_tokens` small. Autocomplete suggestions don't need to be 1,000 tokens long. 20–100 is usually right. The default may be larger; set it explicitly.

Use Flash, not Pro. Autocomplete rewards speed and cost over marginal quality. Pro is overkill for completing the next line of code.

Stop sequences cut off output early. If you're completing within a function and don't want the model to write past the closing brace, set `stop=["}", "\n\n"]` to terminate generation at logical boundaries.

Failure Modes

A few things to watch for with FIM:

Hallucinated context. The model sometimes generates content that contradicts the suffix — defining a variable in the middle that conflicts with how the suffix uses it, for example. Watch for this in your evaluation.

Suffix-ignoring completions. If the model doesn't quite "see" the suffix (or the suffix is far from the gap), it may generate as if there's no suffix at all. For long files, the suffix immediately after the gap carries the most weight.

Style mismatch. The middle generation should match the surrounding code's style — naming conventions, indentation, patterns. When it doesn't, it's usually a sign the surrounding context isn't long enough, or the model is defaulting to its own preferences.

Truncation at `max_tokens`. Hit `max_tokens` mid-statement and you get incomplete output. Set `max_tokens` generously for typical completions but use stop sequences to catch logical endings.

What FIM Is Used For in the Real World

In practice, FIM powers:

- **Editor autocomplete** in tools like Cursor, Continue.dev, and similar AI-augmented IDEs
- **CLI code completion** for shell-based AI tools
- **Document templating** in workflow automation tools
- **Code modernization** where one function needs to be rewritten in surrounding-code-style
- **Scaffolding generators** that fill in structured templates

If you're building any of these, FIM is your primitive. If you're building something else, standard chat completion is probably what you want.

The next chapter starts Part IV — deeper into reasoning and thinking-mode workflows, including the patterns that make agent loops cost-effective.

Part IV: Reasoning and Thinking Mode

"The unexamined answer is not worth giving."

— With apologies to Socrates

Chapter 16: When to Use Thinking Mode

Chapter 7 introduced thinking mode in the web app — a per-call toggle that produces explicit chain-of-thought reasoning before the final answer. This chapter goes deeper, from a developer's angle: the cost calculus, the patterns where thinking mode genuinely earns its higher token spend, and the strategies for using it selectively rather than universally.

For API workloads at scale, the thinking-mode decision is meaningful in dollars. Done right, you pay for reasoning where it matters and skip it where it doesn't. Done badly — always on, or never used — you either burn substantial cost on calls that didn't need it or accept lower quality on calls that did.

The Cost of Thinking

Thinking mode generates a chain-of-thought reasoning trace before the final answer. The trace counts as output tokens, which (a) cost more per token than input tokens, and (b) increase latency proportional to length.

A concrete example. Suppose a typical non-thinking call generates: - 800 input tokens (prompt) - 200 output tokens (response)

The same call with thinking mode might generate: - 800 input tokens (same prompt) - 1,500 reasoning tokens + 200 final-answer tokens = 1,700 output tokens

The output tokens grew 8.5×. At V4 Flash output rates of ~$0.28/M, the cost per call grew from ~$0.000056 to ~$0.000476 — about an order of magnitude.

Latency scales similarly. A 200-token response that took 2 seconds becomes a 1,700-token response that takes 12–15 seconds.

For one-off calls, these numbers are noise. For production workloads at scale they're real money and real user-perceived slowness. Thinking mode has to earn its cost.

When Thinking Mode Earns Its Cost

The patterns where thinking mode reliably improves answer quality:

Multi-step Reasoning

Problems that require getting A right, then B based on A, then C based on B. Without thinking mode, the model often skips or conflates steps. With it, the explicit reasoning forces sequential treatment.

Examples: - "If the company's revenue grew 15% YoY in 2024 and they project 20% growth in 2025, what would 2025 revenue be if 2023 revenue was $4M?" - "Given the migration constraints, what's the safe sequence for cutting over each service?" - "Walk through what happens when this code executes with input [X]."

Mathematical Work

Algebra, probability, statistics, calculus, formal logic. The model is far more reliable at math when it works through the problem step by step.

The classic test: ask any reasonably complex probability question with thinking off and on. The non-thinking answer is often confidently wrong; the thinking answer is usually right (and the reasoning is auditable).

Code That Has to Actually Work

For non-trivial code generation — implementations with edge cases, multi-part refactors, debugging gnarly issues — thinking mode dramatically improves correctness. The reasoning step makes the model consider what could go wrong, plan the structure, and check its own logic.

For boilerplate (CRUD endpoints, simple data transformations, well-known patterns), thinking mode is overkill.

Complex Planning and Strategy

Multi-stakeholder plans, sequenced approaches, anything where the answer is a coherent plan rather than a discrete fact. Reasoning helps the model weigh dependencies, trade-offs, and implementation order.

Verification and Critique

Asking the model to evaluate an existing answer — "is this argument sound?", "what's wrong with this code?", "where would this plan fail?" — thinking mode produces much more careful evaluation.

When Thinking Mode Is Wasteful

The opposite end: workloads where thinking mode just adds cost without improving outcomes.

Simple Factual Retrieval

"What's the boiling point of water?" "Who wrote Pride and Prejudice?" "What's the capital of France?" The model knows. Reasoning doesn't help.

Stylistic Writing Tasks

Rewriting an email, polishing prose, generating creative variations, translating between languages. These reward feel and craft, not deliberate reasoning. Thinking mode tends to produce stilted output.

Open-Ended Brainstorming

When you want quantity and variety ("give me 30 product names"), thinking mode actually narrows the output by working one path more thoroughly. Direct generation gives more diversity.

Pure Pattern Generation

Generating examples, producing templates, filling in known structures. The model has the pattern; it just needs to apply it.

Trivial Classification

"Is this email spam?" "Is this sentiment positive?" Single-class classification with clear training signal works fine at direct response. Thinking mode adds latency without lifting accuracy.

Per-Call Thinking Decisions

In production, thinking mode is best applied per-call rather than as a global default. The mechanics are simple — set or omit the parameter on each request:

```
def call_with_optional_thinking(messages, use_thinking=False):
    extra = {"thinking": {"type": "enabled"}} if use_thinking else {}
    return client.chat.completions.create(
        model="deepseek-v4-pro",
        messages=messages,
        extra_body=extra,
    )
```

Your application logic decides per-call whether `use_thinking=True` is appropriate. A few decision strategies:

Heuristic-Based

Match keywords or features in the request to the patterns where thinking helps:

```
THINKING_SIGNALS = [
    "explain why", "explain how", "walk through", "step by step",
    "calculate", "what if", "compare", "analyze", "design", "debug",
]

def needs_thinking(message):
    low = message.lower()
    return any(s in low for s in THINKING_SIGNALS) or has_math_notation(message)
```

Classifier-Based

Use a small Flash call to classify whether the next call needs thinking:

```
def classify_for_thinking(message):
    classification = client.chat.completions.create(
        model="deepseek-v4-flash",
        messages=[
            {"role": "system", "content": "Does this request require step-by-ste
p reasoning to answer well? Respond only 'yes' or 'no'."},
            {"role": "user", "content": message},
        ],
        max_tokens=10,
    )
```

```
    return classification.choices[0].message.content.strip().lower().startswith
("yes")
```

The classification call costs little but routes more accurately than heuristics for nuanced cases.

User-Specified

Where users can signal intent, expose thinking mode as a UI control:

```
# In your API, accept a parameter
def chat_endpoint(messages, careful_mode=False):
    return call_with_optional_thinking(messages, use_thinking=careful_mode)
```

Power users opt in when they need it. Works well for technical or analytical applications where users understand the trade-off.

Cascade

Try non-thinking first; escalate to thinking only if the answer fails a quality check:

```
def cascade_thinking(messages, quality_check):
    response = call_with_optional_thinking(messages, use_thinking=False)
    if quality_check(response):
        return response
    # Escalate
    return call_with_optional_thinking(messages, use_thinking=True)
```

The quality check might look for specific failure patterns (the answer says "I don't know"), validate against expected output structure, or use a separate Flash call to evaluate the answer.

Reading and Using Reasoning Output

When thinking mode produces a reasoning trace, it appears in a designated field of the response. The field is named `reasoning_content`; the principle is that the trace is separate from the final `content`.

For application code, three options for the reasoning:

Discard it. For workflows where users only need the final answer, the reasoning is invisible and you can drop it after the response.

Display it. For technical or analytical UIs where users want to see the work, render it alongside the answer (or behind a "show reasoning" disclosure).

Audit on it. For high-stakes workflows, log the reasoning for later review. The trace is invaluable when answers turn out to be wrong — you can see where the model went off, which often suggests how to prompt better next time.

The reasoning is not a private mental state. It's just generated text, calibrated through training to produce useful intermediate work. Treat it as a working artifact that's sometimes more valuable than the final answer.

Combining Thinking Mode with Tool Calling

Thinking mode and tool calling complement each other well. The reasoning helps the model decide when to call which tool, what arguments to pass, and how to interpret the results.

```
response = client.chat.completions.create(
    model="deepseek-v4-pro",
    messages=messages,
    tools=tools,
    extra_body={"thinking": {"type": "enabled"}},
)
```

For agent workflows where the model has multiple tools and needs to plan a sequence of calls, thinking mode dramatically improves tool-selection quality. The cost is real (more tokens per call); the quality difference on complex agent tasks is also real.

Chapter 18 goes deeper on agent patterns.

A Cost-Aware Thinking Strategy

Putting it together — a production strategy that respects cost and quality:

Default off. Most calls don't need thinking mode. Don't enable it by default.

Heuristic on for clearly reasoning-bound calls. Math problems, explicit "step by step" requests, multi-step planning prompts.

Cascade for quality-critical workloads. Try fast first; escalate to thinking when the cheap answer fails a quality check.

User-controlled for technical applications. Where users understand the trade-off, expose the toggle.

Always on for genuinely hard tasks. Some workloads are inherently reasoning-dependent — formal verification, complex coding agents, careful analysis. For these, always-on is the right default.

Always off for stylistic and creative work. Don't add reasoning where it doesn't help.

Log reasoning traces for high-stakes calls. When the work matters, the audit trail is worth the storage cost.

A Worked Example

Concrete numbers on a representative production workload.

Suppose you run a customer support agent handling 10,000 tickets per day. Without thinking, each ticket takes one Flash call averaging: - 1,500 input tokens - 300 output tokens

Daily cost (without caching for simplicity): - Input: 10,000 × 1,500 × $0.14 / 1M = $2.10 - Output: 10,000 × 300 × $0.28 / 1M = $0.84 - **Daily: $2.94**

Turn thinking mode on for every call (output grows to ~2,000 tokens including reasoning):

- Output: 10,000 × 2,000 × $0.28 / 1M = $5.60
- **Daily: $7.70** — nearly 3× the cost

Now suppose 20% of tickets are genuinely complex and 80% are routine. With heuristic routing:

- Routine (80%): 8,000 × 1,500 input + 8,000 × 300 output = baseline cost (~$2.35)
- Complex (20%): 2,000 × 1,500 input + 2,000 × 2,000 output = ~$1.54
- **Daily: ~$3.89**

Selective thinking adds about a third to the cost, applied only where it improves outcomes. Universal thinking would have nearly tripled the bill while only meaningfully helping a fifth of the calls.

That's the discipline. Thinking mode is powerful. It's also expensive. Apply it deliberately.

What Comes Next

The next chapter takes thinking mode into the domain where it delivers the largest quality lift: mathematical and scientific reasoning. We'll cover Prover-V2 for formal mathematics, the patterns for math-heavy chat workflows, and the practical limits of what reasoning models can and can't do.

Chapter 17: Mathematical and Scientific Reasoning

The single domain where reasoning models — DeepSeek's V4 with thinking mode, R1, and the dedicated Prover-V2 — produce the largest quality lift over standard generation is mathematical and scientific work. The same prompt that produces confidently-wrong answers from a non-thinking model often produces correct, auditable, step-by-step solutions from a reasoning model.

This chapter covers the patterns for math and science workloads: when V4 with thinking mode is the right tool, when DeepSeek-Prover-V2 is, the limits of what reasoning models can verify, and the workflows for mathematicians, researchers, students, and analysts whose work involves quantitative reasoning.

Where Reasoning Models Excel

The math and science workloads where reasoning models produce dramatic quality lifts:

Probability and statistics. Setting up problems, applying the right distribution, computing expected values, understanding conditional probability. The reasoning trace makes the model's approach auditable — you can see whether it set up the problem correctly before checking the arithmetic.

Algebra and calculus. Symbolic manipulation, equation solving, derivatives, integrals. Step-by-step working catches errors that direct response would propagate.

Discrete math and combinatorics. Counting problems, graph reasoning, recursion analysis — the kind of work where the answer depends on enumerating cases carefully.

Logical reasoning. Formal logic, conditional reasoning, proofs by contradiction. Reasoning mode doesn't turn the model into a complete proof assistant, but it lifts capability substantially over direct generation.

Scientific computation. Multi-step physics problems (forces, equilibrium, energy conservation), chemistry stoichiometry, biology problems requiring careful application of concepts.

Empirical analysis. Working through what a dataset might show, what statistical tests would apply, what causal interpretation is appropriate.

For all of these, the difference between thinking on and thinking off is often the difference between "useful tool" and "confidently wrong noise."

Where Reasoning Models Struggle

Even with thinking mode, some math/science workloads have real failure modes:

Long arithmetic chains. A problem requiring 50+ multiplication steps will eventually have arithmetic errors. The model isn't a calculator; it generates plausible-looking numbers. For workloads requiring exact arithmetic, use code execution (tool calling, code interpreter) rather than asking the model to compute.

Domain-specific notation. Highly specialized notation in advanced math or physics may not be well represented in training data. The model can sometimes work with novel notation if you define it explicitly, but expect quality drops at the cutting edges.

Truly novel problems. The model is good at problems with analogues in training data. Genuinely novel problems — open research questions, problems requiring new techniques — are beyond what current models reliably solve.

Verification of complex proofs. The model can produce proofs that look correct but contain subtle errors only a formal proof assistant would catch. For high-stakes proof verification, that's what Prover-V2 (below) and external tools like Lean are for.

Geometric reasoning. Spatial and visual reasoning is harder for language models than algebraic reasoning. Problems requiring strong spatial intuition often produce errors even with thinking mode.

The discipline that maps onto these failure modes is awareness — knowing when to trust the model, when to verify with external tools, and when to do the math yourself.

The Verification Workflow

A practical pattern for serious math work with DeepSeek:

1. **Set up the problem with thinking mode on.** Let the model produce its reasoning. Read the setup — is the problem framed correctly?
2. **Read the reasoning critically.** Does each step follow from the previous? Are there logical jumps that need scrutiny?
3. **Verify the final arithmetic.** Use a calculator, spreadsheet, or code execution to confirm the numbers.
4. **For results that matter, cross-check.** Run the problem through a separate tool (Wolfram Alpha, a CAS like Mathematica or SymPy, or careful manual work), or have the model approach the problem from a different angle and confirm the answer matches.
5. **For high-stakes work, get human review.** The model is a powerful assistant, not a replacement for domain expertise.

The reasoning trace is what makes this workflow practical. Without it, you'd just have an answer to verify in isolation. With it, you can audit the approach.

Code Execution for Exact Computation

For any math involving more than trivial arithmetic, the right pattern is to have the model **set up the problem** and **execute the math in code** rather than relying on the model to compute.

In the web app, that's the code interpreter feature. In the API, it's tool calling with a math/code-execution tool. The pattern:

```
tools = [{
    "type": "function",
    "function": {
        "name": "execute_python",
        "description": "Execute Python code and return the result. Use this for
any non-trivial arithmetic.",
        "parameters": {
            "type": "object",
            "properties": {
```

```
                "code": {"type": "string"},
            },
            "required": ["code"],
        },
    },
}]
```

The model writes Python that does the computation; your tool runs it; the result feeds back to the model for interpretation.

This combination — model reasoning for setup and approach, code execution for exact arithmetic — is dramatically more reliable than either alone. For production analytical workloads, it's the right architecture.

DeepSeek-Prover-V2 for Formal Mathematics

For users at the cutting edge of formal mathematics — Lean theorem proving, formal verification, automated theorem proving — DeepSeek released **Prover-V2**, a specialized model trained for proof generation and verification within the Lean theorem prover framework.

Prover-V2 is genuinely state of the art at its niche. It can:

- Generate Lean proofs for stated theorems
- Verify proofs you've written (within Lean's verification framework)
- Suggest proof strategies for difficult conjectures
- Translate informal mathematical arguments into formal Lean proofs

The audience is small but real: working mathematicians, formal methods researchers, formal verification teams (smart contract verification, critical software), and the growing population of computer scientists treating mathematics formally.

Most users will never need Prover-V2. For the audience that does, it's the strongest open-weights tool currently available for formal mathematical reasoning.

Chapter 27 covers Prover-V2 in more depth.

Math-Heavy Workflows for Different Audiences

A few specific patterns by user type:

For Researchers

Use V4 with thinking mode for: - Working through problems before writing them up - Checking your own reasoning on a derivation - Generating draft proofs for theorems you're trying to formalize - Exploring whether a conjecture might hold by checking small cases

For high-stakes verification (formal proofs, peer-reviewed work), pair the reasoning model with formal tools (Lean, Coq, etc.). The reasoning model is excellent at first pass; formal verification is what makes it trustworthy.

For Students

Use V4 with thinking mode for: - Working through homework problems (with the discipline of doing them yourself first, then comparing approaches) - Understanding solution strategies (not memorizing answers) - Getting unstuck when you're not sure how to approach a problem - Verifying your own solutions

The pedagogical risk is real: students who substitute the reasoning model for their own engagement develop fewer skills. The students who use it well treat it as a tutor that explains approaches, not an answer machine.

For Analysts

Use V4 with thinking mode + code execution for: - Statistical analyses where you need to set up the right test - Financial calculations where arithmetic precision matters - Quantitative reasoning where the model handles the framing and code handles the math - Sanity-checking results from your existing tools

The combination of reasoning model + code execution is particularly powerful for analyst work — careful problem setup AND exact computation.

For Engineers

Use V4 with thinking mode for: - Performance analysis (Big O reasoning, throughput calculations) - Capacity planning (traffic load, queue dynamics) -

Reliability analysis (failure mode reasoning, MTBF calculations) - Algorithm correctness arguments

For systems where the math is load-bearing for production decisions, treat the model's output as a starting point for verification, not a final answer.

A Worked Example: A Probability Question End-to-End

A concrete walkthrough. Suppose you're working through a problem:

> *"A clinical trial has a 70% probability of detecting a true effect. If the true effect exists in 30% of cases studied, what's the probability that an observed positive result reflects a true effect rather than a false positive?"*

This is Bayes' theorem applied to clinical trial reasoning. With thinking mode on, V4 produces something like:

> *Let me set up Bayes' theorem here. P(true effect | positive result) = P(positive | true effect) × P(true effect) / P(positive) We know: - P(positive | true effect) = 0.70 (sensitivity / power) - P(true effect) = 0.30 (prior probability) - We need P(positive), which expands using total probability... ... P(true effect | positive) = 0.21 / (0.21 + 0.07 × P(false positive rate))*
>
> *I notice the problem doesn't specify the false positive rate. I'll assume the standard alpha = 0.05. ... Final: approximately 86% probability that the positive result reflects a true effect.*

Notice what the reasoning gave you:

1. **The framework** — Bayes' theorem, applied correctly
2. **The variable identification** — what's known, what's needed
3. **The assumption** — the model flagged a missing piece of data (false positive rate) and made a reasonable assumption
4. **The intermediate values** — auditable arithmetic
5. **The final answer** — with the underlying calculation visible

A non-thinking response would have produced a number with no work shown. You'd have to trust it. With the reasoning, you can verify each step (or feed the same problem into a CAS or spreadsheet to confirm the arithmetic).

With code execution available, the model could write the computation in Python and you'd get an exact numerical answer with zero arithmetic errors.

What Reasoning Models Don't Replace

Even at their best, reasoning models don't replace:

- **Domain expertise.** Knowing which problem to solve, which assumptions are reasonable, which results are surprising.
- **Formal verification.** For high-stakes math (proofs in published research, verification of critical systems), formal tools are required. Reasoning models help you get there; they don't replace the verification.
- **Calculators and computational tools.** For exact arithmetic at scale, use code execution or specialized tools, not the model.
- **Peer review.** For published work, human expert review of the reasoning is non-negotiable.

What they do replace, or at least dramatically accelerate:

- **First-pass problem setup**
- **Working through routine derivations**
- **Generating candidate solutions for verification**
- **Pedagogical tutoring**
- **Sanity-checking your own work**
- **Exploring whether a hard problem is tractable**

That's the right framing: reasoning models as a powerful first stage in mathematical workflows, with verification and domain expertise as the necessary subsequent stages.

The next chapter covers the most ambitious application of reasoning models — agent workflows, where the model uses tools, makes multi-step decisions, and navigates complex tasks autonomously.

Chapter 18: Agent Workflows

Agent workflows are the most ambitious thing you can build with a language model. The model uses tools, makes decisions, observes the results, makes more decisions, and navigates a multi-step task toward an outcome — without you scripting each step in advance.

For some workloads, agents are transformative. They handle the kind of multi-step structured work that traditionally required human attention or brittle automation scripts. For others, they're an expensive, slow, error-prone solution to problems that don't need them.

This chapter covers when to build agents, how reasoning models change agent design, the standard patterns (plan-then-execute, reflection, critique loops), and the cost-control disciplines that keep agent workloads tractable.

What an Agent Is

A working definition: an **agent** is a language model in a loop. The model receives an objective, decides what to do, takes an action (typically by calling a tool), observes the result, and decides what's next. The loop continues until the model decides the objective is complete, or some external limit (step count, cost, time) terminates it.

The basic agent loop from Chapter 13:

```
def run_agent(objective, max_steps=20):
    messages = [{"role": "user", "content": objective}]
    for step in range(max_steps):
        response = client.chat.completions.create(
            model="deepseek-v4-pro",
            messages=messages,
            tools=available_tools,
            extra_body={"thinking": {"type": "enabled"}},
        )
        msg = response.choices[0].message
        messages.append(msg)
        if not msg.tool_calls:
            return msg.content  # Agent reports done
        for tc in msg.tool_calls:
```

```
            result = execute_tool(tc)
            messages.append({"role": "tool", "tool_call_id": tc.id, "content": j
son.dumps(result)})
    raise RuntimeError("Agent didn't terminate")
```

That's the canonical shape. Real production agents add layers — planning, validation, error recovery — but the core loop is what makes them agents rather than chains.

When Agents Are the Right Architecture

Agents earn their cost when:

The task requires multiple tool calls in unknown sequence. If you knew the exact sequence in advance, you'd hard-code it. Agents are for when the sequence depends on intermediate results.

The task involves observation and adaptation. The agent does something, looks at the result, decides whether to keep going, change approach, or stop. That's the heart of why agents matter.

The objective is high-level. "Investigate this incident and tell me what happened," "research this competitive landscape," "fix the failing tests in this PR" — these are objectives that decompose into many concrete steps the agent figures out.

Errors should be recoverable. Agents handle errors by reasoning about them and trying differently. For workflows where each step must succeed deterministically, agents are the wrong model.

When agents are the wrong architecture:

The task is single-step. "Classify this support ticket" doesn't need an agent. Just call the model.

The sequence is fixed and known. "Fetch user, validate, send email" is a workflow. Hard-code it. You'll get faster, cheaper, more reliable execution than an agent provides.

Errors must produce specific structured failures. Agents are messy with certain kinds of errors; deterministic workflows handle them cleanly.

Cost or latency is tightly constrained. Agents are expensive (many model calls per task) and slow (many round-trips). For high-volume, low-latency workloads, the economics don't work.

How Reasoning Models Change Agent Design

Before reasoning models, building reliable agents was hard. The model would call tools, but its decisions about which tool, when, and what to do with results were often wrong in subtle ways. Reliability scaled poorly with task complexity — a 3-step agent might be 90% reliable, a 6-step agent 70%, a 12-step agent 30%.

Reasoning models changed this. With thinking mode, the model produces explicit reasoning about its decisions. It thinks about which tool to use, considers alternatives, plans sequences, and reflects on results — all visible in the reasoning trace. The result: meaningfully more reliable agent behavior on the same tasks.

The implications for agent design:

Thinking mode is approximately mandatory for serious agents. The cost is real (more output tokens per call), but the quality difference on multi-step tasks is dramatic. Disable it only for the simplest agents.

V4 Pro outperforms V4 Flash for agent work. The capability gap shows up most clearly in multi-step planning. For agents, the model upgrade often more than pays for itself in reduced loop iterations.

Reasoning traces are valuable for debugging. When an agent fails, the trace shows where it went wrong. Invaluable for iterating on agent design.

Standard Agent Patterns

A few patterns have emerged as reliable across agent applications:

Plan-Then-Execute

The agent first produces an explicit plan — a sequence of intended steps — then executes it. If a step fails or returns unexpected results, the agent re-plans and continues.

```
def plan_then_execute(objective):
    # Phase 1: Plan
```

```
    plan_response = client.chat.completions.create(
        model="deepseek-v4-pro",
        messages=[
            {"role": "system", "content": "Produce a numbered step-by-step plan
to accomplish the objective. Each step should be a specific action."},
            {"role": "user", "content": objective},
        ],
        extra_body={"thinking": {"type": "enabled"}},
    )
    plan = plan_response.choices[0].message.content

    # Phase 2: Execute (using standard agent loop with the plan as context)
    messages = [
        {"role": "user", "content": f"Objective: {objective}\n\nPlan:\n{plan}\n\
nExecute the plan. Re-plan if a step fails or returns unexpected results."}
    ]
    return run_agent_loop(messages)
```

Plan-then-execute is more reliable than pure react-style loops when the steps are reasonably predictable. For objectives where the next step truly depends on intermediate results, react-style loops are still appropriate.

Reflection

After producing a result, the agent explicitly evaluates its own work and revises if needed:

```
def with_reflection(objective):
    initial = run_agent(objective)

    reflection = client.chat.completions.create(
        model="deepseek-v4-pro",
        messages=[
            {"role": "system", "content": "Review the agent's output critically.
 Identify any errors, gaps, or improvements. If the work is good, say 'PASSED'.
Otherwise, describe what needs to be fixed."},
            {"role": "user", "content": f"Objective: {objective}\n\nAgent output:
\n{initial}"},
        ],
        extra_body={"thinking": {"type": "enabled"}},
    )
    review = reflection.choices[0].message.content

    if "PASSED" in review:
```

```
        return initial

    # Revise
    revision = run_agent(f"Improve this output based on the review.\n\nObjective:
{objective}\n\nInitial output:\n{initial}\n\nReview:\n{review}")
    return revision
```

Reflection adds cost (an extra evaluation call) but catches a meaningful share of low-quality outputs that would otherwise ship.

Cost-Controlled Branches

When you have multiple plausible approaches and don't know which will work, kick off cheap exploratory branches in parallel and pick the best:

```
import asyncio

async def explore_branches(objective, n_branches=3):
    # Each branch: V4 Flash quick attempt
    branches = await asyncio.gather(*[
        run_agent_async(objective, model="deepseek-v4-flash", max_steps=5)
        for _ in range(n_branches)
    ])

    # Evaluate which produced the best result
    best = await select_best(branches)

    # Promote the winner with V4 Pro for the final, careful version
    return await run_agent_async(
        f"{objective}\n\nApproach: {best}",
        model="deepseek-v4-pro",
    )
```

Works for tasks where exploration is cheap relative to final execution — research, investigation, design exploration.

Cascade Agents

Use Flash for the bulk of agent work; escalate to Pro for the hard sub-tasks:

```
def cascade_agent(objective, max_steps=20):
    messages = [{"role": "user", "content": objective}]

    for step in range(max_steps):
        # Try Flash first
```

```
        response = call_model("deepseek-v4-flash", messages, tools)

        if not is_satisfactory(response):
            # Escalate to Pro for this step
            response = call_model("deepseek-v4-pro", messages, tools)

        # Process and continue loop
        ...
```

The savings are real: Flash handles routine steps cheaply; Pro is reserved for the hard ones.

Cost Control: The Non-Negotiables

Agent workloads can run away. A misbehaving loop, even briefly, can rack up real costs in minutes. A few non-negotiable controls for production agents:

Hard Step Limits

Every agent loop needs a maximum step count. Pick a value generous enough for typical tasks (10–20 for most), strict enough to terminate runaway loops.

```
MAX_AGENT_STEPS = 20

def safe_agent(objective):
    for step in range(MAX_AGENT_STEPS):
        # ... agent loop body
        pass
    raise RuntimeError(f"Agent failed to complete within {MAX_AGENT_STEPS} steps
")
```

The step limit isn't a quality bound — it's a runaway-prevention bound. A task that genuinely needs 50 steps is a task you shouldn't have handed to an agent without breaking it down.

Per-Task Cost Caps

Beyond step counts, track cumulative token usage per agent task and abort if it exceeds a budget:

```
class AgentBudget:
    def __init__(self, max_cents=50):
        self.spent_cents = 0
```

```
        self.max_cents = max_cents

    def charge(self, response):
        cost = calculate_cost(response.usage)
        self.spent_cents += cost
        if self.spent_cents > self.max_cents:
            raise BudgetExceeded(f"Agent exceeded {self.max_cents}¢ budget at {s
elf.spent_cents}¢")
```

For a customer-facing agent serving real users, a per-task cap might be a dollar or two. For an internal research agent, $10–20 may be reasonable. The right number depends on workload and business value, but having a cap matters.

Per-Tenant Rate Limits

For multi-user systems, throttle agent invocations per tenant. A single user shouldn't be able to launch a hundred agents at once. Per-user concurrent-agent limits prevent this.

Timeouts on Tool Calls

External tool calls (especially network calls) can hang. Wrap them in timeouts so a slow API doesn't keep an agent loop alive longer than intended.

Circuit Breakers

If an agent repeatedly fails (too many tool errors, repeated invalid responses), break the loop and surface the failure rather than retrying. Repeated failures are usually a symptom of something fundamentally wrong, not a transient glitch worth retrying through.

Observability for Agents

Agents are harder to debug than single calls. A few observability practices help.

Log every model call. Inputs, outputs, tools used, costs, latencies. When something goes wrong, you'll need the trail.

Persist reasoning traces. When the agent goes off the rails, the reasoning is where you'll see it. Don't discard.

Tag agent runs with task identifiers. Group all calls within a single agent invocation under a task ID for trace analysis.

Track success rates over time. A drop in success rate is often the first signal of regression — model behavior change, tool degradation, prompt drift.

Monitor cost per task. Sudden cost jumps usually mean an agent is looping more than expected.

For production agent systems, observability isn't optional. The complexity of multi-step model behavior makes it impossible to debug from logs alone without systematic tracing.

A Worked Example: A Research Agent

Putting the patterns together — a research agent that handles open-ended investigation:

```
TOOLS = [
    web_search_tool,
    fetch_url_tool,
    summarize_tool,
    save_finding_tool,
]

SYSTEM_PROMPT = """You are a research agent. Your job is to investigate the user
's question
and produce a structured report.

Workflow:
1. Plan your research approach (no tool calls in this step - just think)
2. Execute searches and fetch sources
3. Summarize each source
4. Save key findings as you go
5. When you have sufficient findings, produce a final report and stop

Stop when you have enough information to answer well, even if you could keep sea
rching.
Don't over-research."""

def research_agent(question, budget_cents=200, max_steps=15):
    messages = [
        {"role": "system", "content": SYSTEM_PROMPT},
```

```
        {"role": "user", "content": question},
    ]
    budget = AgentBudget(max_cents=budget_cents)

    for step in range(max_steps):
        response = client.chat.completions.create(
            model="deepseek-v4-pro",
            messages=messages,
            tools=TOOLS,
            extra_body={"thinking": {"type": "enabled"}},
        )
        budget.charge(response)

        msg = response.choices[0].message
        messages.append(msg)

        if not msg.tool_calls:
            return msg.content  # Agent done

        for tc in msg.tool_calls:
            try:
                result = execute_tool(tc)
            except Exception as e:
                result = {"error": str(e)}
            messages.append({
                "role": "tool",
                "tool_call_id": tc.id,
                "content": json.dumps(result),
            })

    raise RuntimeError("Agent didn't terminate within step limit")
```

A complete research agent in about 40 lines. It:

- Has clear instructions (system prompt)
- Uses thinking mode for careful tool selection
- Has a step limit
- Has a cost budget
- Handles tool errors gracefully (returns them to the model rather than crashing)
- Stops when the model decides it's done

For most agent applications, this is the foundation. Specific tools, prompts, and constraints customize it for the use case.

When Agents Should Hand Off to Humans

One pattern worth calling out: many agent systems are most useful when they hand off to humans at the right moment rather than trying to complete everything autonomously.

Examples: - A customer support agent handles routine queries; escalates complex ones to a human - A research agent gathers and synthesizes; the human reviews and decides - A coding agent proposes changes; the human reviews and merges - An incident-investigation agent investigates; the human takes action

Designing the handoff is part of agent design. A clean handoff — with all the context the human needs — is much more valuable than an agent that tries (and fails) to do everything.

What Closes Part IV

That closes Part IV. You now know when thinking mode is worth its cost, how to apply it to mathematical and scientific work, and how to build agents that use it well.

Part V is for users self-hosting DeepSeek — running the open-weight models on their own infrastructure rather than via the API. If you're API-only, skip ahead to Part VI for the audience-specific chapters.

Part V: Self-Hosting and Open Weights

"Owning the means of inference."

— A working slogan

Chapter 19: The Open-Source Story

DeepSeek's most consequential decision wasn't the model architecture or the pricing — it was releasing the weights under a permissive open-source license. A frontier-class model that anyone could download, fine-tune, deploy, and modify reset what was possible at the cutting edge of AI.

For most users, this matters indirectly: open weights keep the API cheap (DeepSeek can't charge a premium for capabilities competitors can replicate), keep DeepSeek honest (you can audit the model in ways API-only providers don't permit), and create a downstream ecosystem of fine-tunes, distillations, and derived work that wasn't possible with closed-weight models.

For users who want to run the models themselves — for privacy, for cost at scale, for capabilities not available via the API, or just for the joy of having frontier AI on hardware they control — the open weights are the core product.

This chapter covers the open-source story in detail: the licensing, what you actually get when you download weights, when self-hosting makes sense versus when you should just use the API, and the hardware reality of running V4 versus distilled models.

MIT Licensing and What It Permits

DeepSeek's models are released under the **MIT license**, one of the most permissive open-source licenses in software. In plain language:

You can use the model for any purpose. Personal, commercial, research, educational. No restrictions.

You can modify the model. Fine-tune it, distill it, ablate it, combine it with other models, build derivative works.

You can redistribute the model. Pass it along to others, host it as a service, embed it in products you sell.

You can do all of this without telling DeepSeek. No registration, no royalties, no commercial-use restrictions.

The only requirement is preserving the license notice. Distributions need to include the MIT license text and copyright attribution.

Compare this to most other major AI providers:

- **OpenAI** doesn't release weights at all. The model is API-only.
- **Anthropic** doesn't release weights for Claude.
- **Google** has released some smaller Gemma models under open weights but not the full Gemini line.
- **Meta** releases Llama under a license that's permissive but has commercial-use restrictions for the largest deployments.

DeepSeek's MIT license is genuinely the most open of any frontier-tier model lab. Whatever you want to do with the weights, you can do.

What "Open Weights" Actually Means

When DeepSeek releases a model, what you can download:

The trained weights. The numerical parameters that define the model — typically tens to hundreds of gigabytes of binary data, distributed via Hugging Face.

The model architecture specification. Code (typically Python with PyTorch or similar) describing the model structure so the weights can be loaded and used.

Tokenizer files. The vocabulary and tokenization rules the model was trained with.

Configuration files. Hyperparameters, special tokens, generation defaults.

Documentation. Usually a model card describing training, capabilities, limitations, and recommended use.

What you typically **don't** get:

- The training data (proprietary, often legally complex)
- The full training code (parts may be released, but training a frontier model from scratch isn't typically reproducible from public release)

- Fine-tuning data for the model's chat behavior (sometimes released, often not)

For inference — running the model — you have everything you need. For training a new model from scratch using DeepSeek's exact recipe, you don't.

Compliance and Commercial Use

A practical note for teams considering commercial use of open-weight DeepSeek:

MIT license use is straightforward. Your legal team will recognize the MIT license; it's been used in open-source software for decades. No AI-specific licensing complexity.

Trademark stays with DeepSeek. You can use the weights freely, but calling your fine-tuned variant "DeepSeek-Whatever" risks trademark issues. Use distinct names for your derivatives.

You're responsible for outputs. The model can generate problematic content; the license doesn't shift that responsibility to DeepSeek. You're the one running it.

Export controls may apply. Depending on jurisdiction, AI models may be subject to export controls. The MIT license doesn't exempt you from export law.

For most commercial users, none of these are real obstacles. The MIT license makes commercial deployment about as easy as it gets in AI.

The Hardware Reality

The single biggest factor in whether self-hosting makes sense for you is hardware. Frontier models are huge.

V4 Pro: The Full-Size Flagship

DeepSeek-V4-Pro is **1.6 trillion total parameters**. At standard BF16 precision, that's roughly 3.2 TB of weights. With 8-bit quantization (FP8 or INT8), about 1.6 TB. With aggressive 4-bit quantization, about 800 GB.

To run inference at reasonable throughput, you need GPU memory holding the active weights plus working memory for activations. For V4 Pro at FP8, that's **8+**

high-end data-center GPUs (e.g., H100 or H200 nodes) at minimum — and that's tight. For comfortable production deployment, expect a multi-node cluster.

Capital cost: an 8x H100 node was running $250–300K as of early 2026. Cloud rental is $20–50 per GPU-hour, so $160–400/hour for an 8-GPU node.

For a workload that runs continuously, the math is roughly: - API costs (V4 Pro at discounted rates): ~$0.50/M output tokens × your token volume - Self-hosted: ~$200K capital + ~$3K/month operational, regardless of token volume

The break-even is high. For most applications, the API is dramatically cheaper.

V4 Flash: The Cost-Efficient Sibling

V4 Flash at 284 billion parameters is more tractable: roughly 570 GB at BF16, 285 GB at FP8, 142 GB at INT4. Single-node deployment on 4–8 GPUs is feasible.

Still expensive — a multi-GPU node with enough VRAM for V4 Flash at reasonable precision runs $100K+ capital, $50–150/hour cloud.

Distilled Models

This is where self-hosting becomes practical. The distilled models — 1.5B, 7B, 8B, 14B, 32B, 70B parameters — are sized for hardware most teams already have:

- **1.5B**: Runs on edge devices, browsers, mobile. CPU inference is slow but possible.
- **7B/8B**: Consumer GPU (8–16 GB VRAM) handles it comfortably. Modern laptops with discrete graphics work.
- **14B**: 24 GB VRAM (RTX 4090, A4000, similar) handles it well at quantized precision.
- **32B**: 48–80 GB VRAM (single A100, dual 4090s, etc.) handles it.
- **70B**: 80+ GB VRAM (A100 80GB, H100, multi-GPU consumer setup) handles it.

For most teams that genuinely want to self-host, the distilled lineup is what they actually deploy. Capability is meaningfully below V4 Pro but appropriate for many workloads — and the hardware is within budget.

When Self-Hosting Makes Sense

Self-hosting is the right answer when:

Privacy requirements are absolute. Some workloads (regulated industries, sensitive corporate data, classified work) genuinely cannot use external APIs. For these, self-hosting on infrastructure you control is the only option.

Latency requirements are tight. API calls round-trip across the internet. For workloads requiring low-millisecond inference (high-frequency trading, real-time signal processing), self-hosting on local hardware can deliver lower latency.

API access is unreliable in your context. Some jurisdictions have unreliable internet, regulatory blocks on external APIs, or compliance constraints that make API-only architectures impractical.

You're at a scale where break-even flips. For very high token volumes (tens of billions per month), self-hosted costs amortize favorably. The fixed costs of GPU infrastructure spread across enough requests to undercut API pricing.

You need capabilities only available in self-hosted form. Custom fine-tuning, modified prompting paths, integration with internal training data, research access to model internals.

You have hardware you might as well use. Teams with existing GPU clusters (often inherited from other ML work) may have an effective marginal cost close to zero.

When the API Is the Right Answer

For most teams, most of the time, the API is the right answer:

Cost. Below very high volumes, API costs are dramatically lower than the all-in cost of self-hosting (hardware, ops, maintenance, on-call).

Operational simplicity. Self-hosting frontier-class models is real ops work. GPU drivers, CUDA versions, inference framework upgrades, model storage, monitoring, scaling — it's a real responsibility.

Capability access. The API is updated continuously. Self-hosted models are snapshots. Falling behind on capabilities is a real cost.

Engineering attention. Time spent operating self-hosted infrastructure is time not spent building your actual product.

Reliability. DeepSeek's API has its own SLA. Your self-hosted deployment is your responsibility, on your hardware, with your alerts.

For most production applications: use the API. Reserve self-hosting for the cases where the specific advantages above genuinely apply.

The Hybrid Approach

Many teams end up with a hybrid: API for most workloads, self-hosted for specific cases.

Common patterns:

Self-host distilled models for sensitive workloads, API for everything else. A 14B distilled model handling internal-only queries, V4 Pro via API for customer-facing work.

Self-host for batch processing, API for interactive. Run a self-hosted cluster for overnight batch jobs (where utilization is high); use the API for interactive workloads (where utilization is bursty).

Self-host for training/fine-tuning, API for production. Use your own infrastructure to fine-tune custom variants; deploy them on infrastructure you control or via specialized providers; use the standard API for everything else.

The hybrid approach gets you the privacy and capability benefits of self-hosting where they matter, with the cost and operational benefits of the API where they don't.

What's Coming in This Part

The next three chapters cover the practical mechanics of self-hosting at different scales:

- **Chapter 20: Running DeepSeek Locally with Ollama** – for laptop-scale deployment of distilled models, the entry point most users will start with

- **Chapter 21: Production Deployment with vLLM and BentoML** — for serious production deployment, the patterns and tools that make multi-GPU serving tractable
- **Chapter 22: The Distilled Models** — the lineup, the trade-offs, and when each size is appropriate

If self-hosting isn't relevant to your work, skip ahead to Part VI for the audience-specific chapters. The principles in Parts I–IV transfer to API-only use; Part V is specifically for the self-hosted path.

For users staying in: the self-hosting story is more accessible than it sounds. With distilled models and good tooling (Ollama in particular), running real DeepSeek capability on your own laptop is a 30-minute setup. The next chapter covers exactly that.

Chapter 20: Running DeepSeek Locally with Ollama

Ollama is the easiest path from "I don't have DeepSeek running locally" to "I have DeepSeek running locally." One install command, one pull command, and you have a working DeepSeek deployment on your laptop or workstation, accessible via a clean local API.

This chapter walks through Ollama setup, model selection for your hardware, running a basic chat session, exposing the model via an API, and the performance characteristics you should expect.

If you're a developer who wants to experiment with self-hosted DeepSeek without committing to production infrastructure — or a researcher who wants reproducible local inference — Ollama is the right starting point.

What Ollama Is

Ollama is an open-source tool for running language models locally. It handles model downloading, format conversion, GPU/CPU dispatch, and quantization decisions, and exposes a clean local API for programmatic access. Think of it as Docker for local LLMs — the abstraction over the messy details of model serving.

A few key properties:

Cross-platform. Native installs for macOS, Linux, and Windows (with WSL support).

Quantization-aware. Handles GGUF format models (the de facto standard for local inference) with various quantization levels (Q4, Q5, Q8). You can usually run larger models than the naive math suggests.

Local API. Exposes an OpenAI-compatible API on `localhost:11434`. Your application code doesn't have to change to point at it.

Model registry. Pre-packaged models from a public registry, including DeepSeek's distilled models and many community variants.

Hardware-aware. Detects your GPU (NVIDIA, AMD, Apple Silicon) and dispatches inference appropriately. Falls back to CPU if no GPU is available.

Installing Ollama

The install is one command per platform.

macOS: Download the installer from ollama.com and run it. Or via Homebrew: `brew install ollama`.

Linux: `curl -fsSL https://ollama.com/install.sh | sh` — installs Ollama and a systemd service.

Windows: Download the Windows installer from ollama.com.

After install, the `ollama` command should be on your PATH. Verify:

```
$ ollama --version
```

On Linux and Windows, Ollama runs as a background service (started automatically after install). On macOS, run `ollama serve` to start it (or it auto-starts when you run `ollama run`).

Pulling a DeepSeek Model

Ollama maintains a model registry. To download and run a DeepSeek model:

```
$ ollama pull deepseek-r1:7b
```

This downloads the 7B distilled DeepSeek-R1 model. The download is several gigabytes — a few minutes on a typical broadband connection.

A few model variants you might pull, depending on hardware:

- `deepseek-r1:1.5b` — smallest, runs on basically anything
- `deepseek-r1:7b` — works on consumer GPUs (8–16 GB VRAM) or modern Apple Silicon
- `deepseek-r1:14b` — needs ~16 GB VRAM at default quantization
- `deepseek-r1:32b` — needs 24+ GB VRAM, or M2 Max/M3 Max territory on Apple Silicon
- `deepseek-r1:70b` — needs 48+ GB VRAM (A100, dual 4090s, M2 Ultra)

- `deepseek-v3` and variants — larger; check the registry for current sizes

The exact model names in the Ollama registry shift as new releases ship. Run `ollama list` to see what you've already pulled, or browse ollama.com/library for current options.

Picking the Right Model for Your Hardware

A practical sizing guide:

Laptop with 8 GB RAM, no discrete GPU: stick to 1.5B. Runs on CPU; slow but usable.

Laptop with 16 GB RAM, integrated graphics or low-end discrete GPU: 7B works. Modest speed.

Modern Apple Silicon (M1/M2/M3 Pro or Max): 7B works fast, 14B works at reasonable speed, 32B works on M2/M3 Max with 64+ GB unified memory.

Workstation with NVIDIA GPU (8–16 GB VRAM): 7B fast, 14B at quantized precision.

Workstation with NVIDIA GPU (24 GB VRAM, e.g., RTX 4090): 14B fast, 32B at quantized precision.

Workstation with NVIDIA GPU (48+ GB VRAM): 32B fast, 70B at quantized precision.

Multi-GPU workstation: 70B fast, larger models possible with appropriate setup.

The principle: pick the largest model that fits comfortably in memory at your preferred quantization. Bigger models within your hardware limits beat smaller models leaving compute headroom unused.

Running a Chat Session

Once a model is pulled, start chatting from the terminal:

```
$ ollama run deepseek-r1:7b
>>> What's the capital of France?
The capital of France is Paris.
```

```
>>> Why is it the capital?
Paris became the capital of France through a long historical process...
```

The model loads on first use (a few seconds for smaller models, longer for larger), then responds to each prompt. Type `/bye` or Ctrl+D to exit.

A few useful interactive commands:

- `/?` — list all available commands
- `/clear` — clear the current context (start fresh in the same session)
- `/show modelfile` — see the model's configuration
- `/save <name>` — save the current session as a model variant
- `/load <name>` — load a previously saved session

For one-shot prompts without entering interactive mode:

```
$ ollama run deepseek-r1:7b "What's the capital of France?"
The capital of France is Paris.
```

The HTTP API

The real power of Ollama is the local HTTP API. Once Ollama is running, you can call it programmatically just like the cloud.

Native Ollama API

```
import requests

response = requests.post(
    "http://localhost:11434/api/generate",
    json={
        "model": "deepseek-r1:7b",
        "prompt": "Explain RAG in two paragraphs.",
        "stream": False,
    },
)
print(response.json()["response"])
```

OpenAI-Compatible API

For applications already using the OpenAI client, Ollama exposes an OpenAI-compatible endpoint at `localhost:11434/v1`.

```python
from openai import OpenAI

client = OpenAI(
    base_url="http://localhost:11434/v1",
    api_key="ollama",  # any non-empty string works locally
)

response = client.chat.completions.create(
    model="deepseek-r1:7b",
    messages=[
        {"role": "user", "content": "Explain RAG in two paragraphs."},
    ],
)
print(response.choices[0].message.content)
```

This is the magic moment: code that targeted the DeepSeek API now works against your locally-hosted model with one line changed (the `base_url`). Switching between cloud and local becomes trivial.

Streaming

Streaming works identically:

```python
stream = client.chat.completions.create(
    model="deepseek-r1:7b",
    messages=[{"role": "user", "content": "Explain RAG in two paragraphs."}],
    stream=True,
)
for chunk in stream:
    delta = chunk.choices[0].delta.content
    if delta:
        print(delta, end="", flush=True)
```

Performance Expectations

Local inference performance varies wildly by hardware. Rough numbers as of early 2026, all for 7B distilled DeepSeek at Q4 quantization:

- **CPU only (modern laptop)**: 2–5 tokens/second. Usable but slow.
- **Modern Apple Silicon (M2 Pro)**: 30–50 tokens/second. Snappy.
- **Consumer NVIDIA GPU (RTX 3060, 12 GB)**: 30–60 tokens/second. Snappy.

- **Workstation NVIDIA GPU (RTX 4090, 24 GB)**: 80–150 tokens/second. Very fast.
- **Apple M3 Max**: 60–90 tokens/second on 7B; still usable for 32B at 15–25 tokens/second.

For comparison, the cloud API typically streams at 50–150 tokens/second for V4 Flash and 30–80 tokens/second for V4 Pro, with the network adding 100–300 ms of round-trip latency per call.

For interactive use on capable local hardware, the local experience often **feels** faster than the cloud despite lower raw throughput — no network latency. For batch workloads, the cloud is usually faster overall.

Customizing Models with Modelfiles

Ollama supports Modelfiles — Dockerfile-like definitions that customize a base model. A common use:

```
# my-deepseek-coder.modelfile
FROM deepseek-r1:7b

PARAMETER temperature 0.2
PARAMETER top_p 0.9
PARAMETER num_ctx 8192

SYSTEM """You are an expert Python developer. Always produce production-ready co
de with type hints and docstrings. Be concise."""
```

Then build and use it:

```
$ ollama create my-deepseek-coder -f my-deepseek-coder.modelfile
$ ollama run my-deepseek-coder
```

The custom variant inherits the base model's weights but applies your system prompt and parameters automatically. Useful for role-specific or task-specific variants without writing the system prompt every time.

Common Issues and Fixes

A few common Ollama issues and their fixes:

Out of memory. The model is too big for your hardware. Pick a smaller variant or use more aggressive quantization (look for `:q4` or `:q3` variants).

Very slow inference. Likely running on CPU when you expected GPU. Check `nvidia-smi` (Linux/Windows) or Activity Monitor (Mac) to verify GPU utilization. May need updated GPU drivers or CUDA.

Model not found. Either it hasn't been pulled (`ollama pull <model>`) or the name is misspelled (check `ollama list`).

Context exceeded. The default context window is small (often 4K tokens). For longer contexts, set `num_ctx` higher in a Modelfile or pass it as a parameter. Memory cost scales with context length, so don't set it higher than you need.

API not responding. The Ollama service may not be running. On Linux: `systemctl status ollama`. On macOS: ensure `ollama serve` is running.

When to Outgrow Ollama

Ollama is excellent for laptops, workstations, and single-user dev environments. Where it falls short:

Multi-user serving. Ollama is designed for single-user-at-a-time inference. To serve multiple concurrent users from one server, you want a proper inference framework.

Production reliability. Ollama is fine for development; production needs more — health checks, monitoring, graceful restart.

Multi-GPU deployment. Ollama supports basic GPU dispatch but isn't optimized for multi-GPU tensor parallelism the way production inference frameworks are.

High-throughput batching. For workloads doing many concurrent requests, batched inference frameworks dramatically outperform Ollama's request-at-a-time serving.

For all of these, the next chapter is where to go: vLLM and BentoML for production deployment.

What Ollama Enables

For most readers, Ollama is sufficient and excellent:

- **Local privacy** — code, documents, and conversations stay on your machine
- **Zero API cost** — once you've got the hardware, marginal cost is electricity
- **Offline use** — works without internet
- **Experimentation** — try variants, swap models, tinker with prompts
- **Development environment** — build applications against local Ollama, swap in cloud API for production

For developers in particular, a working local LLM completely changes what kinds of experiments feel possible. The setup investment is small (under an hour); the long-term value is large.

The next chapter is for users moving past Ollama into production deployment — multi-GPU serving, high throughput, real reliability.

Chapter 21: Production Deployment with vLLM and BentoML

Ollama is for laptops. Production deployments — multi-GPU serving, high throughput, real reliability — need different tools.

The two most common production stacks for self-hosted DeepSeek as of early 2026 are **vLLM** (a high-throughput inference engine) and **BentoML** (a production-grade serving and deployment framework). They're often paired: vLLM as the inference engine, BentoML as the wrapper that handles deployment, scaling, and operations.

This chapter covers both, when to use each (or both), the deployment patterns that work, and the operational realities of running self-hosted DeepSeek at scale.

Why Production Deployment Differs

Production deployment of self-hosted models has requirements that Ollama doesn't address:

Throughput. A production deployment serves many concurrent requests. Throughput per GPU matters more than per-request latency. Throughput-oriented inference engines (vLLM, TGI, TensorRT-LLM) batch requests together to maximize GPU utilization.

Reliability. Production deployments need health checks, graceful restart, monitoring, and automated recovery. The inference framework alone doesn't handle this; you need a serving layer (BentoML, Ray Serve, custom infra) on top.

Multi-GPU and multi-node. Frontier models often don't fit on a single GPU; they need tensor parallelism across multiple GPUs (or pipeline parallelism across multiple nodes). Production engines handle this transparently; Ollama largely doesn't.

Quantization and optimization. Production deployments often run quantized models for cost, sometimes with hardware-specific optimizations (FP8 on Hopper GPUs, INT8 on older hardware). Production engines support these.

Observability. Per-request metrics, GPU utilization, queue depth, error rates. Critical for running production reliably; missing from simpler tools.

vLLM: The High-Throughput Inference Engine

vLLM is an open-source inference engine built for high-throughput LLM serving. Its key innovation is **PagedAttention**, a KV-cache management technique that dramatically improves memory efficiency and enables much higher request batching than naive serving.

The headline numbers: vLLM typically delivers **2–10× the throughput of naive Hugging Face Transformers serving** on the same hardware, with comparable per-request latency.

Installing vLLM

```
pip install vllm
```

vLLM requires a CUDA-compatible NVIDIA GPU (or AMD ROCm, with separate setup). Most production deployments run on Linux with NVIDIA GPUs.

Serving DeepSeek with vLLM

vLLM ships with a built-in OpenAI-compatible API server. To serve a DeepSeek model:

```
python -m vllm.entrypoints.openai.api_server \
    --model deepseek-ai/DeepSeek-R1-Distill-Qwen-7B \
    --port 8000 \
    --tensor-parallel-size 1
```

For larger models requiring multiple GPUs, increase `tensor-parallel-size`:

```
python -m vllm.entrypoints.openai.api_server \
    --model deepseek-ai/DeepSeek-V3 \
    --port 8000 \
    --tensor-parallel-size 8 \
    --quantization fp8
```

This launches a server on port 8000 that accepts OpenAI-compatible requests. Application code targeting `http://your-server:8000/v1` works without modification.

Key vLLM Parameters

The parameters that matter most for production:

`--tensor-parallel-size` — Number of GPUs to shard the model across. Required for models that don't fit on a single GPU.

`--max-model-len` — Maximum context length. Capping it saves memory; sizing it to your workload matters.

`--quantization` — Reduces memory and increases throughput at some quality cost. Options include `awq`, `gptq`, `fp8`, `int8`. For DeepSeek models, FP8 on Hopper GPUs is the typical sweet spot.

`--max-num-seqs` — How many concurrent requests can be batched together. Higher values increase throughput; too high causes memory pressure.

`--gpu-memory-utilization` — Fraction of GPU memory vLLM is allowed to use. Default 0.9; lower if you have other GPU workloads sharing the device.

Expect to spend time tuning these for your specific workload and hardware. The defaults are reasonable starting points but rarely optimal.

Performance Characteristics

For a deployment of, say, DeepSeek-R1-Distill-Llama-70B on an 8x H100 node with FP8 quantization, expect roughly:

- **Time to first token**: 100–300 ms
- **Per-request throughput**: 30–80 tokens/second
- **Aggregate throughput**: 1,500–5,000 tokens/second across all concurrent requests
- **Concurrent request handling**: 50–200 simultaneously

These numbers vary substantially with prompt length, quantization, model variant, and hardware specifics. They're roughly the range a well-tuned production deployment achieves.

BentoML: The Production Serving Framework

vLLM is an inference engine. It serves requests but doesn't handle the broader concerns of production — model packaging, deployment to Kubernetes, autoscaling, observability, multi-environment promotion.

BentoML is a production-grade serving framework that wraps inference engines (vLLM included) with these capabilities. Pairing the two is a common production pattern.

What BentoML Adds

BentoML provides:

Standardized model packaging. Define your service in Python; BentoML packages it with all dependencies for reproducible deployment.

Deployment automation. Ship to Kubernetes, BentoCloud (BentoML's hosted platform), or other targets without hand-crafting deployment manifests.

Multi-model orchestration. Serve multiple models from one service on shared infrastructure — useful for cost-efficient deployments.

Production monitoring. Built-in metrics, logging, and integration with observability stacks.

API gateway features. Authentication, rate limiting, request routing, circuit breakers.

A Minimal BentoML + vLLM Service

```
import bentoml
from openai import OpenAI

@bentoml.service(
    resources={"gpu": 8, "gpu_type": "nvidia-h100"},
    traffic={"timeout": 300},
)
class DeepSeekService:
    def __init__(self):
        # vLLM serves on a side-car port; client points at it
        self.client = OpenAI(
            base_url="http://localhost:8000/v1",
```

```
            api_key="local",
        )

    @bentoml.api
    def chat(self, messages: list, model: str = "deepseek-v3") -> dict:
        response = self.client.chat.completions.create(
            model=model,
            messages=messages,
        )
        return {"content": response.choices[0].message.content}
```

The service deploys with one command:

```
bentoml deploy
```

Behind the scenes, BentoML builds a container, provisions the GPU resources, runs vLLM as the inference engine, and exposes your `chat` API on a managed endpoint.

For teams comfortable with Python and Kubernetes, BentoML cuts substantial ops overhead versus rolling your own deployment from scratch.

Multi-GPU and Multi-Node Deployment

Models too large for a single GPU (V4 Pro, V4 Flash, large distilled models) need tensor parallelism across multiple GPUs. Models too large for a single node need pipeline parallelism across multiple nodes.

vLLM handles both with the `tensor-parallel-size` and `pipeline-parallel-size` parameters. The math:

- **Tensor parallelism**: shards each layer across N GPUs. Lower latency, requires high-bandwidth interconnect (NVLink, InfiniBand).
- **Pipeline parallelism**: places different layers on different GPUs/nodes. Higher latency, lower bandwidth requirement.

For DeepSeek's larger models, the typical pattern is:

- Single-node with 4–8 H100s using tensor parallelism for V4 Flash and large distilled models
- Multi-node with tensor parallelism within each node and pipeline parallelism across nodes for V4 Pro

Multi-node deployment sharply increases operational complexity. For most teams that need V4 Pro capability, the API is cheaper and more reliable than self-hosting.

Quantization Strategies

Quantization reduces model size and increases throughput at some cost to quality. The standard options for DeepSeek deployment:

FP8 (Hopper-only, e.g., H100/H200). Half the memory of BF16 with minimal quality loss. The default for production V4 deployments on modern hardware.

INT8. Half the memory, more quality impact than FP8. Available on older GPUs (A100, A6000) where FP8 isn't supported.

INT4. Quarter the memory, noticeable quality impact. Fits larger models on smaller GPUs at the cost of some capability.

AWQ/GPTQ. Specialized 4-bit quantization schemes with better quality preservation than naive INT4. Common for distilled model deployment on consumer hardware.

For production, the safest path is FP8 on Hopper. Quality nearly indistinguishable from BF16, half the memory, faster inference. On older hardware (A100), INT8 is the typical default.

For small distilled models on consumer GPUs, AWQ or GPTQ at 4-bit is the standard.

Operational Concerns

Running self-hosted DeepSeek in production has real ops requirements. A few areas worth attention:

GPU monitoring. GPU temperature, memory utilization, error rates. NVIDIA's DCGM or simpler tools (`nvidia-smi` exporters) feeding Prometheus.

Request queue depth. When demand exceeds capacity, requests queue. Visibility into queue depth is critical for autoscaling decisions.

Cold starts. Loading a multi-billion-parameter model into GPU memory takes seconds to minutes. Cold-start latency matters; pre-warmed deployments are usually necessary.

Model updates. Deploying a new model version requires a cutover strategy that doesn't drop traffic. Blue-green deployments, canary rollouts, etc.

Cost tracking. Self-hosted infrastructure has costs (hardware, electricity, cloud rental, ops time). Track them alongside the API costs you're avoiding so the trade-off stays visible.

Capacity planning. GPU memory and throughput limits cap how much traffic you can handle. Autoscaling helps but has limits — sometimes you genuinely need to provision more capacity.

These are the same concerns as any production infrastructure, but the GPU-specific dimensions add complexity unique to LLM deployment.

When This All Makes Sense

For a team to invest in production self-hosting infrastructure, the math has to work:

Token volume justifies fixed costs. A reasonable break-even is roughly 50–100M tokens/day. Below that, the API is almost always cheaper. Above it, self-hosting can pay back.

You have ops capacity. Running production GPU infrastructure is real work. Either dedicated infrastructure engineers, an in-house ML platform team, or willingness to use managed services (BentoCloud, etc.) that abstract the ops.

Privacy or capability requirements are real. Either external APIs are off the table (regulated workloads, classified work) or you need capabilities the API doesn't expose (custom fine-tuning, modified model behavior, research access).

Hardware investment fits. Either you have existing GPU infrastructure or the budget for what you'll need.

For everyone else, the API is the right answer. Self-hosting is a real capability with real costs; deploy it when those costs are justified by specific benefits, not because self-hosting feels more sophisticated.

A Practical Path

For teams considering self-hosted production deployment, a sensible path:

Phase 1: Local development with Ollama. Get familiar with the model, validate use cases, write application code against the local API.

Phase 2: Single-node vLLM deployment. Deploy vLLM on a single GPU node (rented from a cloud provider). Run real workloads against it. Measure throughput, latency, quality, cost.

Phase 3: Production deployment with BentoML. Wrap vLLM with proper deployment, monitoring, and scaling. Ship to your production infrastructure.

Phase 4: Optimization and scaling. Tune quantization, batching parameters, autoscaling policies. Scale to multiple nodes as load demands.

This is months of work for a small team. The first two phases land in weeks; phases 3 and 4 take longer because they're production-engineering work, not just model deployment.

The next chapter covers the distilled model lineup — the smaller variants of DeepSeek that make self-hosting practical for most teams. The engines and patterns from this chapter still apply; the difference is which model you serve.

Chapter 22: The Distilled Models

For most teams that genuinely self-host DeepSeek, the distilled models are what they actually deploy. Frontier-class V4 Pro is too large for most infrastructure budgets. V4 Flash is more tractable but still substantial. The distilled lineup — 1.5B to 70B parameters — is sized for hardware most teams already have and delivers capability appropriate for many real workloads.

This chapter covers the distilled lineup in detail: what each model is, what it's good for, the hardware it needs, and how to pick the right size for your use case.

What Distillation Means

In machine learning, **distillation** trains a smaller "student" model to mimic the behavior of a larger "teacher" model. The student doesn't have to figure out how to behave from raw training data; it learns from the teacher's outputs, which compresses substantial capability into a much smaller package.

DeepSeek's distilled models were trained using DeepSeek-R1 (the reasoning model) as the teacher. The students learned to produce R1-style reasoning at smaller sizes. The resulting models aren't as capable as R1 itself, but they're often substantially more capable than other open-weight models of similar size — precisely because they inherited reasoning behavior from a much larger teacher.

The base architecture varies. DeepSeek released distilled variants built on:

- **Llama 3.1** (Meta's open-weight model)
- **Qwen** (Alibaba's open-weight model)

The R1-distilled behavior is the special sauce; the base model determines architectural details (tokenizer, attention pattern, etc.) that affect serving and deployment.

The Lineup

As of early 2026, the standard distilled lineup includes:

DeepSeek-R1-Distill-Qwen-1.5B — Smallest. Runs on essentially any modern hardware including phones (with appropriate inference frameworks).

DeepSeek-R1-Distill-Qwen-7B — The entry point for serious local use. Runs on consumer GPUs.

DeepSeek-R1-Distill-Llama-8B — Llama-base equivalent of the 7B Qwen. Slight architectural differences; pick based on the framework you're already using.

DeepSeek-R1-Distill-Qwen-14B — Meaningfully more capable than 7B/8B. Needs higher-end consumer GPUs.

DeepSeek-R1-Distill-Qwen-32B — A capability sweet spot for many production workloads. Needs workstation-grade or single-data-center GPUs.

DeepSeek-R1-Distill-Llama-70B — Largest distilled variant. Approaches R1's capability on many benchmarks. Needs serious GPU infrastructure.

The exact lineup shifts with new releases. Check Hugging Face for current availability.

Capability vs. Size

A general principle: **larger distilled models are more capable, but the gains are sub-linear.**

The 1.5B is much weaker than the 7B — closer to a coherent assistant than to R1's reasoning.

The 7B and 14B are the "useful local model" tier. They handle most everyday tasks reasonably well — coding, analysis, conversation, basic reasoning.

The 32B is where the distilled models start feeling genuinely capable. Reasoning quality is meaningfully better than the smaller variants. For production workloads where the local model is doing real analytical work, 32B is often where to start.

The 70B is the most capable distilled variant and approaches the R1 teacher on many benchmarks. For workloads requiring maximum capability from a self-hosted deployment, it's what you want — but the hardware requirements are substantial.

Hardware Requirements (Concrete)

Sizing a deployment to your hardware:

1.5B at FP16: ~3 GB VRAM. Runs on basically anything with a GPU, or CPU-only.

1.5B at Q4: ~1 GB VRAM. Phones, browsers, embedded.

7B at FP16: ~14 GB VRAM. Runs on consumer GPUs with 16+ GB.

7B at Q4: ~5 GB VRAM. Runs on most consumer GPUs.

14B at FP16: ~28 GB VRAM. Workstation GPUs (A4000, A5000) or quantized on consumer.

14B at Q4: ~9 GB VRAM. Comfortable on most consumer GPUs.

32B at FP16: ~64 GB VRAM. A100 80GB, dual A6000 (48GB each), Apple Mac Studio Max.

32B at Q4: ~20 GB VRAM. RTX 4090 (24 GB), single workstation GPU.

70B at FP16: ~140 GB VRAM. Multi-GPU production infrastructure.

70B at Q4: ~45 GB VRAM. A100 80GB, dual high-end consumer GPUs.

The sizing rule: pick the largest model that fits comfortably in your VRAM at your preferred quantization, with headroom for context (about 1–2 GB per 1K tokens of context window depending on model).

When to Use Distilled Models Instead of the API

The distilled models earn their keep when:

Privacy or compliance requires local inference. Sending API call data outside your infrastructure isn't acceptable for some workloads. Distilled models on local hardware solve this cleanly.

You have idle GPU capacity. Teams with existing GPU infrastructure (often inherited from other ML work) can deploy distilled models on hardware that's already paid for.

Latency is critical and you have local hardware. Local inference avoids network round-trips. For interactive workflows on capable hardware, distilled models can feel snappier than the cloud API despite lower raw throughput.

The task is well-served by smaller models. Not everything needs frontier capability. For classification, extraction, simple coding, structured generation — 7B or 14B distilled is often entirely sufficient at much lower per-call cost than V4 Pro.

You want to fine-tune for a specific use case. Distilled models are a strong starting point for fine-tuning on domain data. The base capability is solid; specializing for your use case can produce a model that outperforms the API on your specific workload.

When the API is still the right answer:

You need the strongest capability available. No distilled model matches V4 Pro on hard tasks. If you need that capability, use the API.

Operational simplicity matters. Self-hosting is real ops work even with distilled models. Without the engineering capacity, the API is genuinely cheaper once you account for that work.

Variable load. Self-hosting has fixed capacity (your hardware). The API scales elastically. For bursty workloads, the API often wins on cost.

Choosing a Specific Variant

A practical decision tree for picking the right distilled model:

For experimentation/learning what's possible: - Start with **7B** on whatever GPU you have - If it's underwhelming, jump to **14B**

For production deployment of well-understood workloads: - **7B/8B** for classification, extraction, simple Q&A, summarization - **14B** for general assistance, coding, conversation - **32B** for analytical work, complex reasoning, anything the 14B doesn't quite handle - **70B** for tasks at the edge of what self-hosted can do; otherwise consider the API

For edge deployment (phones, browsers, embedded): - **1.5B** at heavy quantization

For fine-tuning starting points: - **7B** if compute is constrained - **14B** for most fine-tuning projects - **32B** for fine-tuning when you have the GPU resources to train on it

Quantization Trade-offs

For each distilled model, you pick a quantization level. The trade-off is memory vs. quality:

FP16 (full precision): maximum quality, maximum memory. Use when you have headroom and quality matters.

FP8: nearly indistinguishable from FP16 on capable hardware (Hopper). Half the memory.

INT8: small quality impact. Half the memory of FP16. Works on older hardware.

Q5: small quality impact. Roughly 5/8 the memory of FP8.

Q4: noticeable quality impact, especially on harder tasks. Roughly half the memory of FP8.

Q3/Q2: significant quality impact. Useful for fitting larger models on small hardware, with the understanding that capability degrades.

For most production use, **FP8 on Hopper or Q4 on consumer hardware** is the standard. The quality loss at Q4 is noticeable on hard reasoning tasks but acceptable for many workloads.

When quality is paramount, run unquantized (FP16) on hardware that can fit it. When running the largest model possible matters more than per-token quality, quantize aggressively.

A Production Deployment Pattern

A common pattern for production self-hosted deployment:

1. **Start with V4 Flash or large distilled (32B/70B) via the API** to validate the workload
2. **Profile real traffic** to understand request shape, throughput requirements, and quality needs

3. **Try smaller distilled models** to see if they're sufficient
4. **Self-host the smallest sufficient model** on infrastructure you control
5. **Maintain API fallback** for spike traffic and edge cases the local model handles poorly

This lets you optimize for cost (smaller is cheaper to host) while preserving capability access (API as fallback). For most workloads where self-hosting makes sense, it's the right shape.

What Distilled Models Don't Replace

A few things even the largest distilled models don't replace:

V4 Pro for highest-capability tasks. The 70B distilled is impressive but doesn't match V4 Pro on the hardest reasoning tasks.

Specialized models. For specific niches (formal math, multimodal vision, OCR), the specialized DeepSeek models (Prover-V2, Janus, OCR) outperform the distilled general models.

Your domain expertise. A 7B model fine-tuned on your specific domain will often outperform a larger general model on your tasks. Domain specialization beats raw scale for narrow problems.

API features. Some API features (caching, hosted thinking-mode infrastructure, managed scaling) aren't replicated when you self-host. You either build them yourself or do without.

What Closes Part V

That closes Part V. You now have the working knowledge to self-host DeepSeek across the full range — laptop experimentation with Ollama, production deployment with vLLM and BentoML, and the distilled lineup that makes self-hosting tractable for most teams.

For the audience that needs this — privacy-constrained workloads, very high token volume, capability that requires self-hosted access, teams with existing GPU infrastructure — Part V is the operational backbone.

For everyone else, the API path covered in Parts I–IV remains the right architecture.

Part VI returns to use cases — applying DeepSeek to specific contexts (developers, researchers, cost-conscious teams) and the comparison case against the major Western providers. If you're trying to figure out where DeepSeek fits in your specific work, Part VI organizes the chapters around your role.

Part VI: DeepSeek for Your World

"The best tool is the one that fits the work."

— A working principle

Chapter 23: DeepSeek for Developers

For developers, DeepSeek is one of the most consequential additions to the AI tooling landscape in years. Not because it does fundamentally new things — the major commercial APIs all do similar work — but because the cost economics let you do far more with AI in your engineering workflow than was previously practical. Token-budget anxiety stops being a constant background concern. Workflows that were too expensive to consider become routine.

This chapter covers the developer-specific patterns: where DeepSeek fits in coding workflows, how it integrates with the AI-augmented IDEs and CLI tools that have become standard, the production-engineering patterns for code-focused services, and where DeepSeek wins on quality-per-dollar over competing options.

Where DeepSeek Wins for Developers

Specific developer use cases where DeepSeek's quality-per-dollar advantage shows up most clearly:

High-volume code review. Running automated review on every PR is expensive when each review costs $0.50–$2 of GPT-5.5 calls. With V4 Flash or Pro, the same coverage costs cents per PR. Teams that couldn't justify universal AI review can now deploy it.

Code generation at scale. Developer tools that generate code (scaffolding, migrations, test fixtures, documentation) that would have been cost-prohibitive on premium APIs become economically viable.

Pair-programming patterns. Sustained AI assistance during development burns API calls. With DeepSeek, your daily coding budget might be a few dollars instead of fifty.

CI/CD integration. AI-driven test generation, change-impact analysis, security scanning — workloads that run on every commit benefit enormously from the cost reduction.

Internal developer agents. Self-built agents that automate engineering tasks (incident triage, on-call runbooks, log analysis) become viable at organizations that couldn't justify the API cost on Western providers.

Education and learning. AI-augmented tutoring for engineers learning new technologies. Cost matters for both individuals (out-of-pocket) and organizations supporting many developers.

The common thread: workloads where AI capability is valuable but the per-call cost on premium APIs would have made the workload economically marginal or fully impractical.

Coding Quality: V4 Pro vs. The Competition

The harder question: how does DeepSeek's coding capability actually compare to the leading alternatives?

As of early 2026, the rough picture:

Claude Opus 4.7 remains the consensus best general-purpose coding model. Strongest at large refactors, complex debugging, and code that requires deep architectural understanding.

GPT-5.5 is competitive across most coding tasks. Particularly strong at idiomatic code in popular languages and at generating well-structured code from natural language descriptions.

DeepSeek-V4 Pro with thinking mode is competitive with both on most coding workloads. Slightly behind on the very hardest tasks (complex multi-file refactors, debugging gnarly issues in unfamiliar codebases), comparable on most everyday work, sometimes ahead on specific niches (open-source library code, architectures it has more training exposure to).

DeepSeek-V4 Flash is meaningfully below the others on hard tasks but very competitive for routine coding work — and at a tiny fraction of the cost.

The practical implication: for most coding workloads where you're using AI as an assistant, DeepSeek is competitive enough to be the default. For the hardest problems where every quality point matters and cost is irrelevant, Claude Opus or GPT-5.5 may still be worth the premium. For everything in between — most production developer workloads — DeepSeek's cost advantage is decisive.

Integration with Developer Tools

DeepSeek's API is OpenAI-compatible, so it works with most developer tools that support OpenAI's API. The list as of early 2026 includes:

Cursor

The AI-native code editor (covered in detail in *Cursor: The Complete Guide*) supports custom OpenAI-compatible endpoints. Configure Cursor to use DeepSeek's API:

```
{
  "name": "DeepSeek V4 Pro",
  "url": "https://api.deepseek.com/v1",
  "model": "deepseek-v4-pro",
  "api_key": "sk-..."
}
```

Cursor will route to DeepSeek for the calls you make through its features. Useful for keeping Cursor's per-credit pricing under control while still using a strong model.

Claude Code

Anthropic's terminal-based coding agent uses Anthropic's models by default, but the underlying patterns work with OpenAI-compatible endpoints if you configure them. Some teams run Claude Code-style workflows against DeepSeek for cost reduction.

GitHub Copilot

Copilot's commercial product uses OpenAI under the hood and doesn't expose a model swap. But Copilot-style pair programming patterns can be replicated using DeepSeek — see the open-source equivalents (Continue.dev, Cline) below.

Continue.dev

An open-source AI code assistant that runs as a VS Code or JetBrains extension. Supports DeepSeek as a first-class configured provider:

```
{
  "models": [
```

```
    {
      "title": "DeepSeek V4 Pro",
      "provider": "openai",
      "model": "deepseek-v4-pro",
      "apiBase": "https://api.deepseek.com/v1",
      "apiKey": "sk-..."
    }
  ]
}
```

A common pattern for cost-conscious developers: Continue.dev + DeepSeek for daily coding, with API costs typically under $5/month for active use.

Aider

A CLI-based AI pair programming tool. Configure for DeepSeek with:

```
aider --model deepseek/deepseek-v4-pro --api-key sk-...
```

For developers who prefer terminal workflows over GUI editors, Aider plus DeepSeek is a strong combination.

OpenCode

A newer CLI coding agent that lists DeepSeek as a first-class supported provider. Out-of-box configuration for V4 Pro.

Custom Tools

For internal developer tools you build yourself, the OpenAI client + DeepSeek base URL pattern from Chapter 4 works exactly as you'd expect. Most developer-tooling SDKs that target OpenAI will work with DeepSeek with a single config change.

Code Generation Patterns

A few patterns that work well for code-generation workloads:

Generate, then Test

Don't trust generated code without verification. The pattern:

1. Model generates code

2. Your tooling runs tests against the generated code
3. If tests fail, feed the failures back to the model
4. Loop until tests pass or step limit hit

```
def code_with_tests(spec, max_attempts=5):
    messages = [{"role": "user", "content": f"Write code for: {spec}"}]
    for attempt in range(max_attempts):
        response = client.chat.completions.create(
            model="deepseek-v4-pro",
            messages=messages,
            extra_body={"thinking": {"type": "enabled"}},
        )
        code = extract_code(response.choices[0].message.content)
        test_result = run_tests(code, spec)
        if test_result.passed:
            return code
        messages.append(response.choices[0].message)
        messages.append({
            "role": "user",
            "content": f"Tests failed: {test_result.failures}. Fix the code.",
        })
    raise RuntimeError("Could not generate passing code in {max_attempts} attemp
ts")
```

The test-driven loop dramatically improves correctness. When the model's first output isn't always right, this pattern lets it self-correct.

Refactor in Context

For refactoring, give the model the full file (or relevant subset) plus the specific change goal:

```
def refactor(file_text, target_function, instruction):
    messages = [
        {
            "role": "user",
            "content": f"""Here's a file:
```

{file_text}

```
Refactor the function `{target_function}` to: {instruction}

Return only the modified function, not the whole file. Preserve the function's e
```

```
xternal interface unless explicitly told otherwise."""
        }
    ]
    response = client.chat.completions.create(
        model="deepseek-v4-pro",
        messages=messages,
        extra_body={"thinking": {"type": "enabled"}},
    )
    return extract_code(response.choices[0].message.content)
```

For non-trivial refactors, V4 Pro with thinking mode is the right model. Flash handles simpler refactors at a fraction of the cost.

Documentation Generation

Generating docstrings, README sections, API documentation:

```
def doc_function(function_code):
    response = client.chat.completions.create(
        model="deepseek-v4-flash",  # Flash is fine for documentation
        messages=[
            {"role": "system", "content": "Generate a docstring for the given fu
nction. Use Google-style format. Be concise and specific."},
            {"role": "user", "content": function_code},
        ],
    )
    return response.choices[0].message.content
```

Documentation generation is the case where Flash plus caching is dramatically cheaper than premium alternatives. Run it across your whole codebase and barely move the needle on monthly spend.

Test Generation

```
def generate_tests(function_code, framework="pytest"):
    response = client.chat.completions.create(
        model="deepseek-v4-pro",  # Pro for tests — they should actually catch b
ugs
        messages=[
            {"role": "system", "content": f"Generate {framework} tests for the g
iven function. Cover happy path, edge cases, and error conditions. Output only t
he test code."},
            {"role": "user", "content": function_code},
        ],
        extra_body={"thinking": {"type": "enabled"}},
```

```
    )
    return extract_code(response.choices[0].message.content)
```

Generated tests need human review before merging — but they're a substantial accelerator for getting test coverage to acceptable levels on existing code.

Production-Engineering Patterns for Code-Focused Services

For developers building services that use DeepSeek for code-related work in production, a few patterns matter:

Cache aggressively. The system prompt for a code service is typically long and stable (instructions, formatting requirements, domain context). Cache hit rates above 90% are achievable; the savings compound across high-volume services.

Route by complexity. Simple code requests (boilerplate generation, documentation) → Flash. Complex requests (architecture decisions, debugging) → Pro with thinking mode. Save 5–10× on the routine portion of traffic.

Use FIM for editor-style autocomplete. As covered in Chapter 15, FIM is the right primitive for low-latency autocomplete suggestions. Don't use chat completion for autocomplete; the latency and quality are wrong for the use case.

Validate generated code. For services that generate code that's then executed (or merged), validation matters. At minimum: parse the generated code (catch syntax errors), run linters, run a small test suite. For higher-stakes generation, full CI before any auto-merge.

Instrument for cost. Code-generation services have wildly variable per-request costs. Instrumenting cost per request (and per user, per feature, per code path) makes optimization possible.

Open-Source Library Code

A specific niche where DeepSeek often outperforms its competitors: code involving popular open-source libraries. The training data appears to include substantial open-source code, and the model handles common framework patterns (Django, FastAPI, React, Next.js, Express, etc.) particularly well.

For workloads heavy in framework-specific code generation, DeepSeek often produces correct, idiomatic code on the first try where competitors require iteration.

Failure Modes to Watch For

Patterns where DeepSeek can fail in characteristically code-related ways:

Hallucinated APIs. Like all LLMs, DeepSeek will sometimes invent API methods or library functions that don't exist. Especially risky for newer libraries where training data is sparse. Always verify imports and API signatures.

Outdated patterns. Training data has a cutoff. Patterns that have evolved since (deprecated APIs, new best practices) may not be reflected. Check generated code against current docs for fast-moving frameworks.

Cross-language confusion. When generating code in less common languages, DeepSeek may import patterns from more common ones. Generated Rust that "feels like" Python or generated Go that has Java idioms — watch for these.

Subtle correctness issues. As with all LLMs, generated code may compile and run but harbor subtle bugs (off-by-one errors, race conditions, incorrect edge-case handling). The "generate then test" pattern catches a meaningful share; human review catches more.

Over-confident security claims. DeepSeek will sometimes generate code that "handles" security concerns superficially without actually being secure. For security-critical code, dedicated review is essential — generative AI doesn't replace security expertise.

A Real PR Workflow

Putting it together, a realistic developer workflow that uses DeepSeek throughout:

Local development. Continue.dev or Cursor configured against V4 Pro for active coding. Daily cost: a few cents to a few dollars.

Test generation. On each commit, a script generates tests for any new functions. Human review before merging. Daily cost: very low (Flash for routine, Pro for complex).

Pre-commit checks. A git hook runs the changed code through DeepSeek for a quick sanity review — catches obvious issues before they hit CI. Daily cost: cents.

PR review. Automated AI review on every PR using V4 Pro with thinking mode. Catches subtle issues, suggests improvements. Per-PR cost: a few cents to ~$0.10 depending on size.

Documentation generation. Periodic batch job updates internal docs based on code changes. Weekly cost: cents.

Total monthly cost for a developer doing all of this: typically under $20. The same pattern on premium APIs lands in the $200–500 range — not impossible, but enough higher to be a budget conversation.

What This Adds Up To

For developers, DeepSeek's contribution to the AI tooling landscape is mostly economic, not capability-based. The capability is competitive; the cost transformation is what changes the math. Workloads that were too expensive to run universally become affordable. Personal AI augmentation becomes financially trivial.

The competitive landscape will keep shifting. Anthropic and OpenAI may close the cost gap. DeepSeek may improve capability further. New entrants will emerge. But the door is open on a different cost regime, and the engineering organizations that adapt earliest will get more out of AI assistance for less money.

The next chapter applies similar treatment to research workloads, where the cost-quality dynamics work out somewhat differently.

Chapter 24: DeepSeek for Researchers

For researchers, DeepSeek changes both economics and possibility. The economics: prompt-experiment pipelines that would burn through grant budgets become tractable. The possibility: open weights enable research approaches that closed-API providers don't permit.

This chapter covers DeepSeek as a research tool — patterns for cost-efficient large-scale prompt experiments, the open-weights story for reproducible research, fine-tuning paths and the hardware they require, and considerations for citing AI tool use in academic work.

Why DeepSeek Matters for Research

A few specific properties of DeepSeek that map onto research workflows:

Open weights = reproducible research. A paper reporting results from V4 Pro can be replicated by anyone with the model weights. Closed-API papers depend on the provider continuing to host the same model version, which they often don't. Open weights are the right substrate for research that has to be reproducible.

Cost-efficient large-scale experiments. Many research workloads involve thousands or millions of model calls — evaluation harnesses, prompt sweeps, ablation studies, scaling experiments. The 10× cost difference vs. premium APIs makes experiments practical that would otherwise require institutional grant funding.

Self-hosted access for sensitive research. Research involving sensitive data (medical, legal, classified) where external API access isn't permitted gains a viable AI option through self-hosted distilled models.

Modifiability. Researchers can fine-tune, ablate, probe, and analyze the model in ways closed-API providers don't permit.

For working researchers, DeepSeek isn't just a cheaper alternative to OpenAI — it enables categories of work the closed-API ecosystem doesn't support.

Cost-Efficient Large-Scale Prompt Experiments

The most common research workload that benefits from DeepSeek economics: large-scale prompt evaluation.

A typical pattern: you have a benchmark of 5,000 problems and want to evaluate model performance across various prompting strategies, conditions, and configurations. Each variation is another 5,000 calls. Five variations across two models is 50,000 calls.

On GPT-5.5, those 50,000 calls might cost $200–500 depending on token sizes. On V4 Flash, the same workload costs $5–15. On a graduate student budget, the difference is between "yes, run the experiment" and "no, that's not in the budget."

A few practical patterns:

The Eval Harness

```
import asyncio
from openai import AsyncOpenAI

client = AsyncOpenAI(
    api_key=os.environ["DEEPSEEK_API_KEY"],
    base_url="https://api.deepseek.com/v1",
)

async def evaluate_problem(problem, prompt_template, model="deepseek-v4-flash"):
    formatted = prompt_template.format(problem=problem.text)
    response = await client.chat.completions.create(
        model=model,
        messages=[{"role": "user", "content": formatted}],
    )
    return {
        "problem_id": problem.id,
        "answer": response.choices[0].message.content,
        "input_tokens": response.usage.prompt_tokens,
        "output_tokens": response.usage.completion_tokens,
    }

async def run_eval(benchmark, prompt_template, model, concurrency=20):
    semaphore = asyncio.Semaphore(concurrency)

    async def bounded_eval(problem):
```

```
        async with semaphore:
            return await evaluate_problem(problem, prompt_template, model)

    return await asyncio.gather(*[bounded_eval(p) for p in benchmark])

# Run
results = asyncio.run(run_eval(my_benchmark, my_prompt, "deepseek-v4-flash"))
```

A few key practices:

Bound concurrency. A semaphore keeps you from blasting the API with thousands of simultaneous requests and tripping rate limits.

Save intermediate results. Evaluations that take hours need periodic checkpointing. Don't lose work to a network blip three hours in.

Cache hits matter at scale. If your prompt template is constant across problems, the prefix is cacheable. A 50,000-call eval with cache hits on a 1,000-token system prompt saves substantially over the same eval without caching.

Log usage per call. Token counts and costs per call let you analyze the experiment's economics afterward — useful for budgeting future experiments and for reporting.

Prompt Sweeps

A common research question: how do small variations in the prompt affect output quality? With DeepSeek's economics, you can test variations exhaustively instead of picking one and hoping.

```
prompt_variations = [
    "Solve this problem step by step: {problem}",
    "Think carefully and solve: {problem}",
    "Problem: {problem}\n\nSolution:",
    # ... dozens more
]

results_by_variation = {}
for variation in prompt_variations:
    results_by_variation[variation] = asyncio.run(
        run_eval(my_benchmark, variation, "deepseek-v4-flash")
    )
```

```
# Now you can analyze: which prompt formulation worked best on which problem typ
es?
```

For 30 prompt variations on a 5,000-problem benchmark, that's 150,000 calls. On Flash with cache-friendly prompts, perhaps $20 of API cost. On premium APIs, 10–20× that. The economics enable comprehensive sweeps that would otherwise have to be cherry-picked.

Cross-Model Evaluation

Evaluating the same workload across V4 Pro, V4 Flash, and various distilled models locally maps the capability-cost frontier for your specific problem.

```
models_to_test = [
    ("deepseek-v4-pro", client),
    ("deepseek-v4-flash", client),
    ("deepseek-r1-distill-qwen-32b", local_client),  # via your local Ollama
    ("deepseek-r1-distill-qwen-7b", local_client),
]

cross_model_results = {}
for model_name, model_client in models_to_test:
    cross_model_results[model_name] = run_eval_against_client(
        benchmark, prompt, model_client, model_name
    )
```

The output is a clear capability frontier: how much capability you lose at each cost point. Critical for deciding what to deploy in production.

Open Weights and Reproducible Research

A long-running tension in AI research: papers that report results from closed-API models are inherently irreproducible. The provider can change the model, retire the version, or alter behavior in ways that mean future researchers can't replicate the paper's results.

Open-weight models like DeepSeek's flip this. A paper that reports V4 Pro results, with the specific version date and parameters, can be reproduced by any researcher with the weights — for as long as those weights remain downloadable (and they're persistent on Hugging Face indefinitely).

For the research community, this matters substantially:

Citation that means something. "We used DeepSeek-V4-Pro-2026-04-24 with thinking mode enabled" is a reproducible specification.

Verification of claims. Other researchers can run the same experiments and check.

Building on prior work. Subsequent papers can use the same model as a baseline instead of guessing.

Long-term research record. Papers from 2026 will still be reproducible in 2030 if the weights remain available.

The practical implication for your own work: when publishing research that uses AI models, prefer open-weight models for the parts where reproducibility matters. Closed-API models are fine for exploratory work or where you just need a strong baseline you can describe; for the experiments that establish the paper's claims, open weights are the right substrate.

Fine-Tuning Paths

For research that requires specialized model behavior — domain-specific reasoning, task-tuned models, calibration to specific evaluation criteria — fine-tuning becomes relevant.

DeepSeek's open weights make fine-tuning practical. The hardware reality varies dramatically by model size:

Distilled 1.5B–8B models. Fine-tunable on a single workstation GPU (RTX 4090 or similar). Memory-efficient training techniques (LoRA, QLoRA) make this feasible without specialized infrastructure.

Distilled 14B–32B models. Need workstation or single-A100 hardware for fine-tuning. Still tractable for individual researchers with reasonable GPU access.

Distilled 70B models. Need multi-GPU infrastructure. Common for well-resourced labs, less accessible for individual researchers without institutional support.

Full V3/V4 models. Genuinely difficult — multi-node infrastructure required. Most fine-tuning at frontier scale is done by labs with dedicated AI infrastructure.

For most research fine-tuning, the distilled models in the 7B–32B range are the practical sweet spot. Capability is meaningful (especially when fine-tuned on domain data), hardware requirements are tractable, and the resulting models can be shared and replicated.

LoRA Fine-Tuning

Low-Rank Adaptation (LoRA) is the dominant fine-tuning approach for resource-efficient training. Instead of updating all model weights, LoRA adds small trainable adapters that capture the fine-tuning differences.

A typical LoRA fine-tuning workflow:

```
from peft import LoraConfig, get_peft_model, TaskType
from transformers import AutoModelForCausalLM, AutoTokenizer
import torch

base_model_name = "deepseek-ai/DeepSeek-R1-Distill-Qwen-7B"
tokenizer = AutoTokenizer.from_pretrained(base_model_name)
model = AutoModelForCausalLM.from_pretrained(
    base_model_name,
    torch_dtype=torch.bfloat16,
    device_map="auto",
)

lora_config = LoraConfig(
    task_type=TaskType.CAUSAL_LM,
    r=16,
    lora_alpha=32,
    lora_dropout=0.1,
    target_modules=["q_proj", "k_proj", "v_proj", "o_proj"],
)

model = get_peft_model(model, lora_config)
# ... rest of training loop
```

Tools like Hugging Face's `transformers` and `peft`, plus higher-level wrappers like `axolotl` and `unsloth`, abstract most of the complexity. For researchers with PyTorch familiarity, fine-tuning a distilled DeepSeek model is a few weeks of work, not months.

Citing DeepSeek in Academic Work

For papers that use DeepSeek models, the citation conventions are still settling. A few practices that hold up:

Cite the model card. DeepSeek publishes model cards on Hugging Face with technical details and recommended citations. Use the format the model card suggests.

Specify version exactly. "DeepSeek-V4-Pro-2026-04-24" is more useful than "DeepSeek-V4-Pro" for future reproducibility.

Document your prompts and parameters. Temperature, top-p, thinking mode, max tokens — all of it should appear somewhere in the paper or supplementary materials. Without these, the reported results aren't fully reproducible even with the model.

Include hardware and software environment. vLLM version, GPU type, quantization — for self-hosted experiments, these affect results.

Acknowledge limitations. The China question and behavioral constraints (Chapter 31) are real and worth acknowledging in papers where they might affect results.

The norms for citing AI tool use in research are evolving fast. Default to over-disclosure rather than under-disclosure: when in doubt, document what you used and how.

Where DeepSeek Falls Short for Research

Honest framing of the limitations:

Frontier capability for the hardest tasks. Some research workloads require the absolute best model available. For these, V4 Pro is competitive but not always leading. Evaluate honestly for your specific use case.

Multimodal limitations. DeepSeek's vision capabilities (Janus) are competitive but not the strongest available. For research heavy in vision tasks, GPT-5.x with vision or Gemini 3.1 Pro may be the better choice.

Specialized domain coverage. Some narrow domains (specific scientific fields, specialized programming languages, particular cultural contexts) may have less training data representation in DeepSeek than in alternatives. Empirical evaluation in your domain is essential.

Hosting reliability for production research applications. DeepSeek's API has reasonable uptime but isn't at the SLA level of major cloud providers. For research applications that need to run reliably for months, build in redundancy.

Updates and deprecation. Models update; older versions get deprecated. For long-running research applications, plan for version management.

A Practical Research Project Pattern

For a researcher starting a new project that involves substantial model use:

Phase 1 — exploration (cheap). Use V4 Flash for initial exploration. Try variations, build intuition for what works.

Phase 2 — design (cheap). Design your experiments based on what you learned. Document the protocols, prompts, and metrics.

Phase 3 — execution (volume varies). Run the experiments. For evaluations across thousands of cases, V4 Flash is your workhorse. For experiments specifically about V4 Pro capability, use Pro.

Phase 4 — verification (selective). Cross-check key findings against alternative models — run a sample of cases against GPT-5.5 or Claude to verify your DeepSeek-based findings replicate. For research claims, multi-model verification strengthens confidence.

Phase 5 — publication. Document everything: model versions, prompts, parameters, hardware. Make experiment code and data available where possible.

For a typical paper using AI models, this pattern (with appropriate adjustments for your specific work) gets you from exploration to publication with costs under control.

From Researcher to Budget-Holder

Chapter 25 looks at DeepSeek through a different lens — cost-conscious teams, where the question isn't just "what can we do" but "what fits the budget." For organizations integrating DeepSeek into existing AI infrastructure, the strategic case comes next.

Chapter 25: DeepSeek for Cost-Conscious Teams

Most engineering organizations using AI in production are paying more for it than they expected. AI features ship; usage grows; the monthly bill from OpenAI or Anthropic creeps upward. Some teams adapt by capping features. Others by raising prices. Most by absorbing the costs and hoping for either provider price drops or capability improvements that justify the spend.

DeepSeek changes the math, but only for teams willing to do the work to integrate it.

This chapter is for the engineering leaders, CTOs, and technical decision-makers who own the AI cost line. It covers the strategic case for DeepSeek as a cost layer beneath existing frontier models, the multi-model routing patterns that make this work in production, the political and operational considerations of adopting a Chinese provider, and the playbook for actually moving traffic.

The Strategic Case

For a team currently spending substantially on OpenAI, Anthropic, or Google's APIs, the strategic question is whether to add DeepSeek as a cost layer. The argument:

The capability gap is small for most workloads. For routine production AI work — chat interfaces, classification, summarization, RAG, coding assistance — DeepSeek is competitive with frontier alternatives. Quality differences exist but are marginal for these workloads.

The cost gap is large. A 5–10× cost reduction is typical when moving suitable workloads from premium APIs to DeepSeek. For organizations spending thousands per month on AI, that's tens of thousands annually. For organizations spending more, larger.

Multi-provider architecture is good engineering anyway. Vendor lock-in to a single AI provider is a strategic risk regardless of cost. Building the abstraction layer that lets you route between providers makes you resilient to price changes, capability deprecations, and provider-specific outages.

The work is bounded. Adding a second provider is a defined engineering project, not an open-ended research effort. For teams with existing AI integration, adding DeepSeek typically takes a few weeks of focused work.

For most organizations spending $5K+/month on AI APIs, the math works in favor of adding DeepSeek. The engineering investment pays back in months, often weeks.

Multi-Model Routing in Production

The architecture for using DeepSeek alongside existing providers:

```
                ┌─────────────────┐
                │  Your Service   │
                └────────┬────────┘
                         │
                         ▼
             ┌────────────────────┐
             │  Routing Layer     │   ← decides which provider
             └─────────┬──────────┘     based on request shape
                       │
          ┌────────────┼──────────────┐
          ▼            ▼              ▼
    ┌──────────┐  ┌──────────┐  ┌──────────┐
    │ DeepSeek │  │  GPT-5.5 │  │ Claude   │
    │  (most)  │  │ (some)   │  │ (some)   │
    └──────────┘  └──────────┘  └──────────┘
```

The routing layer is the new piece. It accepts requests in a unified format and decides which provider handles each one. A few common strategies:

Workload-Based Routing

Each request type is assigned to the provider that handles it best:

```
def route_request(request):
    if request.type == "code_completion":
        return "deepseek-v4-flash"
    elif request.type == "complex_coding":
        return "claude-opus-4.7"
```

```python
    elif request.type == "long_context_analysis":
        return "gemini-3.1-pro"
    elif request.type == "general_chat":
        return "deepseek-v4-flash"
    else:
        return "deepseek-v4-pro"  # default
```

Workload-based routing is the most common pattern. It's understandable, debuggable, and assigns each provider to the workloads where they're strongest.

Cost-Tier Routing

Different user tiers get different providers:

```python
def route_by_tier(user, request):
    if user.tier == "enterprise":
        return "claude-opus-4.7"  # highest capability for paying customers
    elif user.tier == "pro":
        return "deepseek-v4-pro"  # capable, cost-efficient
    else:  # free / trial
        return "deepseek-v4-flash"  # cheapest viable
```

This is a business decision more than a technical one — but it lets you offer differentiated AI quality at different price points without breaking the unit economics of any single tier.

Quality-Cascade Routing

Try the cheap model first; escalate to expensive only when needed:

```python
def cascade_route(request):
    response = call_provider("deepseek-v4-flash", request)
    if quality_check_passes(response):
        return response
    return call_provider("claude-opus-4.7", request)
```

The savings are real: most requests are handled by the cheap model; only the failures escalate. Quality is preserved at the worst-case (premium) bound while average cost drops dramatically.

The hard part is managing the quality check — a separate evaluation step that adds its own cost and latency. For workloads where Flash succeeds 80%+ of the time, cascade routing typically nets 60–80% cost reduction at modest complexity.

A/B-Driven Routing

For teams with established A/B testing infrastructure, routing decisions can be data-driven:

```
def experimental_route(request):
    if should_experiment(request.user_id):
        # 10% of traffic gets DeepSeek for measurement
        provider = "deepseek-v4-pro"
    else:
        provider = "claude-opus-4.7"  # current production default

    response = call_provider(provider, request)
    log_experiment(request, response, provider)
    return response
```

The experiment data tells you whether moving more traffic to DeepSeek would hold quality. Expand the DeepSeek percentage as the data supports it.

Building the Abstraction Layer

The prerequisite for multi-provider routing is an abstraction layer that hides per-provider differences. Most teams end up implementing some version themselves; a few open-source options exist.

A minimal in-house abstraction:

```
class ModelClient:
    def __init__(self, providers):
        self.providers = providers  # dict of provider_name -> client

    def chat_completion(self, model, messages, **kwargs):
        provider, model_id = self._parse_model(model)
        client = self.providers[provider]
        # Translate kwargs to provider-specific format if needed
        return client.chat.completions.create(
            model=model_id,
            messages=messages,
            **self._adapt_kwargs(provider, kwargs),
        )

    def _parse_model(self, model):
        # "deepseek/v4-pro" -> ("deepseek", "deepseek-v4-pro")
        ...
```

```
    def _adapt_kwargs(self, provider, kwargs):
        # Some kwargs differ between providers; normalize
        ...
```

In practice, the abstraction layer evolves to handle:

- **Provider-specific parameter mapping** (e.g., DeepSeek's `extra_body={"thinking": {"type": "enabled"}}` vs. OpenAI's GPT-5.x Thinking model selection)
- **Error handling normalization** so calling code doesn't have to know each provider's error shapes
- **Retry strategies** that account for per-provider rate limits and transient failures
- **Cost tracking** with per-provider attribution
- **Failover** when one provider is degraded

For teams that don't want to build this from scratch, libraries like LiteLLM, OpenRouter, and others provide pre-built provider abstractions. Trade-offs vary; evaluate based on your specific needs.

The Migration Playbook

For a team currently on a single premium provider considering moving substantial traffic to DeepSeek:

Step 1: Baseline measurement. Document current spend, current quality (whatever metrics matter to you), current request mix. You can't measure improvement without a baseline.

Step 2: Build the abstraction layer. Whether in-house or using a library, get to a place where switching providers is a config change, not a code change.

Step 3: Shadow traffic. Send a small percentage of production traffic to DeepSeek in shadow mode (you don't return the response, but you do log it). Compare quality side-by-side with your current provider.

Step 4: Identify safe migration candidates. Use the shadow data to identify request types where DeepSeek matches current quality. Those are your migration candidates.

Step 5: Gradual cutover. Route candidate traffic to DeepSeek with a gradual rollout (1%, 10%, 50%, 100%). Monitor quality and cost throughout.

Step 6: Expand the candidate set. As confidence builds, identify additional workloads to migrate. Repeat the gradual cutover.

Step 7: Maintain the multi-provider architecture. Don't migrate everything to DeepSeek. Keep premium providers for the workloads where they genuinely outperform, and as fallback for DeepSeek outages.

A team starting from scratch on this pattern can typically reach meaningful cost reduction (30–60%) within a quarter. The work is real but well-defined.

Cost Tracking and Attribution

For multi-provider deployments, cost tracking becomes more complex but more important:

Per-provider attribution. How much are you spending on each provider? Daily, weekly, monthly. Trends over time.

Per-workload attribution. Which features or use cases drive the most spend? Optimization opportunities live here.

Per-user attribution. Some users cost dramatically more than others. Knowing this informs pricing and free-tier sizing.

Cost-per-request. Average and distribution. Outliers indicate either expensive workloads or runaway requests.

Migration tracking. As you move workloads to DeepSeek, track the savings. Quantifying the impact justifies continued investment in the abstraction layer.

A simple dashboard showing cost-per-day by provider, plus the same broken down by workload, gives you the visibility needed for ongoing optimization.

The China Question (in This Context)

For teams considering DeepSeek as a cost layer, the political dimension is part of the decision. The honest framing:

DeepSeek operates under Chinese data jurisdiction. Data sent to DeepSeek's API may be subject to Chinese government access requests under Chinese law. For workloads where this matters (government work, regulated industries with specific jurisdictional requirements, certain enterprise customers), DeepSeek may not be acceptable.

The capability comes from Chinese AI research. Some organizations have explicit policies against using AI products from specific countries or organizations. Know your organization's policy.

Self-hosting changes the calculus. For teams using DeepSeek's open weights on infrastructure they control, the data jurisdiction issue largely goes away. The political dimension of supporting Chinese AI research may remain.

Customers may have opinions. Enterprise customers sometimes ask which AI providers you use. Be prepared to answer honestly.

We cover the broader China question in detail in Chapter 31. For cost-conscious teams specifically, the practical guidance: do the diligence to understand your organization's constraints, your customers' constraints, and your regulatory constraints before committing to a DeepSeek-heavy architecture. The cost savings are real; they're not worth violating policy.

When Cost-Conscious Routing Doesn't Make Sense

A few cases where the multi-provider effort isn't worth it:

Very low AI spend. If you're spending under $500/month on AI, the engineering effort to add another provider probably exceeds the savings. Stay simple.

All-or-nothing capability requirements. If your application genuinely needs the strongest model for every request (some research applications, some specialized professional tools), routing doesn't help — you always end up on the premium provider anyway.

Strict compliance requirements. If your requirements exclude DeepSeek entirely, multi-provider routing across acceptable providers (e.g., Claude + Gemini + GPT) still makes sense, but DeepSeek isn't part of the picture.

Resource-constrained engineering teams. Without the engineering bandwidth to maintain a multi-provider architecture properly, sticking with a single provider you operate well beats poorly operating multiple.

For everyone else, multi-provider with DeepSeek as a meaningful component is increasingly the default architecture for AI-native applications.

A Worked Example: 70% Cost Reduction in Six Weeks

A composite example of what this looks like in practice. A SaaS company spending $40K/month on OpenAI for AI features (chat, content generation, classification, embeddings).

Week 1–2: Build abstraction layer. Set up DeepSeek account, test API access, validate that existing application code works against the abstraction.

Week 3: Shadow traffic. Send 10% of production requests to DeepSeek in parallel with OpenAI. Compare outputs on quality metrics.

Week 4: Identify candidates. Classification and routine content generation match OpenAI quality on V4 Flash. Complex reasoning still favors GPT-5.5. Coding assistance favors Claude (separately added to the routing). Migrate classification and routine content generation.

Week 5: Gradual cutover for migration candidates. Monitor quality metrics; no regression. Move to 100% DeepSeek for those workloads.

Week 6: Identify the next round of candidates. Some workloads move to V4 Pro with thinking mode. Embedding workloads stay on OpenAI for now (DeepSeek's embedding capability isn't as mature in this hypothetical).

Result: Monthly AI spend drops from $40K to ~$12K. Quality metrics unchanged or marginally improved on the migrated workloads. Engineering investment: roughly 3 engineer-weeks amortized across the migration.

The savings continue indefinitely. Annual run-rate impact: $336K. The engineering investment paid back within the first month of full migration.

This is what "cost-conscious team using DeepSeek" looks like in practice. The work is real. The economics genuinely change.

Cost is one axis. Chapter 26 takes a different one: the head-to-head against GPT, Claude, and Gemini, focused on where each one is the right choice rather than on cost specifically.

Chapter 26: DeepSeek vs. GPT, Claude, and Gemini

For most teams adopting AI, the question isn't "should we use AI" but "which AI should we use for what." DeepSeek isn't competing in a vacuum — it's competing with GPT-5.4/5.5 from OpenAI, Claude (Sonnet 4.6/Opus 4.7) from Anthropic, and Gemini 3/3.1 Pro from Google. Each has strengths and weaknesses that show up differently across workloads.

This chapter covers the head-to-head: where each provider is genuinely strongest, the operational trade-offs that aren't about raw capability, the pricing landscape, and a framework for deciding which one (or which combination) fits your specific work.

The frame we'll use: not "which is best overall" — that's a meaningless question — but "which is best for what."

The Capability Landscape

An honest characterization of where each provider's models stand as of early 2026.

OpenAI (GPT-5.4, GPT-5.5)

Strengths: - Strongest ecosystem and tooling — most third-party integrations, most documentation, most community knowledge - Reliable across general tasks; few obvious weaknesses - Custom GPTs, Assistants API, and other ecosystem features - Strong agent capabilities with broad tool support - Long history of operational stability

Weaknesses: - Highest pricing in the major-provider tier - Less aggressive on long context than Gemini - Less specialized for any particular niche

When OpenAI is the right choice: - General-purpose AI features where ecosystem matters - Workloads where you want predictable behavior across many sub-tasks - Teams already deep in the OpenAI ecosystem - Custom GPT/Assistants-based applications

Anthropic (Claude Sonnet 4.6, Opus 4.7)

Strengths: - Best-in-class for long-document handling and careful reasoning - Strongest coding capability for complex tasks (especially Opus) - Particularly good at nuanced writing and editing - Cleaner refusals on inappropriate content (fewer false positives) - Constitutional AI training produces measurably more honest outputs in some evaluations

Weaknesses: - Higher pricing than DeepSeek; comparable to OpenAI - Smaller ecosystem than OpenAI - Less aggressive feature shipping cadence

When Claude is the right choice: - Long-document analysis (legal, research, technical specifications) - Complex coding tasks where capability matters more than cost - Writing-heavy workloads where voice and nuance matter - Workloads where careful reasoning and honest uncertainty are valued

Google (Gemini 3, Gemini 3.1 Pro)

Strengths: - Longest context windows widely available (1M+ tokens consistently) - Strong multimodal capability (images, video, audio) - Deep integration with the Google ecosystem (Drive, Docs, Workspace) - Aggressive pricing relative to OpenAI/Anthropic - Gemini 3.1 Pro (powering NotebookLM and the Gemini chat) is competitive across most workloads

Weaknesses: - Less mature ecosystem than OpenAI for enterprise tooling - Behavior less predictable than competitors on some tasks - Tight Google ecosystem integration is also lock-in for non-Google-shop teams

When Gemini is the right choice: - Long-context workloads (massive documents, large codebases) - Multimodal applications (vision, video understanding) - Teams already on Google Workspace - Cost-sensitive workloads where DeepSeek isn't acceptable but premium pricing isn't either

DeepSeek (V4 Pro, V4 Flash, R1, distilled lineup)

Strengths: - 5–10× cost advantage on most workloads vs. OpenAI/Anthropic - Open weights enable self-hosting and reproducible research - V4 with thinking mode is genuinely capable on most workloads - Strong on coding, especially open-source library code - Distilled lineup makes self-hosting practical

Weaknesses: - Behind OpenAI/Anthropic on the very hardest reasoning tasks - Multimodal capability less mature than Gemini or GPT-5.x with vision - China-jurisdiction data handling for hosted API - Politically constrained content (Tiananmen, Taiwan, etc. produce refusals) - Smaller ecosystem of integrations and community tooling

When DeepSeek is the right choice: - Cost-sensitive production workloads at scale - Workloads that need self-hosting (privacy, compliance, latency) - Reproducible research with open-weight requirements - Multi-provider strategies where DeepSeek handles the bulk and others handle exceptions - Coding workloads where the cost difference compounds across volume

Specific Comparisons by Workload

Common workload types and where each provider shines:

General Chat and Q&A

All four providers handle this well. Cost and ecosystem are the main differentiators.

- **Best capability:** Roughly tied; Claude has a slight edge on nuanced/tricky questions
- **Best cost:** DeepSeek (especially V4 Flash)
- **Best for production:** DeepSeek for cost; OpenAI for ecosystem; Claude for quality

Code Generation

- **Best capability:** Claude Opus 4.7 for complex code; tied across the rest
- **Best cost:** DeepSeek V4 Pro
- **Best for production at scale:** DeepSeek for cost-sensitive deployments; Claude for the hardest tasks
- **Best for IDE integration:** Cursor + Claude for premium experience; Cursor + DeepSeek for cost-conscious; Continue.dev + DeepSeek for self-hosted

Long-Document Analysis

- **Best capability:** Claude Opus for careful analysis; Gemini for raw context length
- **Best cost:** Gemini for long-context workloads; DeepSeek competitive
- **Best for production:** Depends on whether you need careful analysis (Claude) or sheer scale (Gemini)

Multimodal (Vision)

- **Best capability:** GPT-5.x with vision and Gemini 3.1 Pro are roughly tied; DeepSeek's Janus is competitive but behind
- **Best for production multimodal:** GPT-5.x or Gemini 3.1 Pro
- **DeepSeek's role:** specific niches (DeepSeek-OCR-2 for documents) where the specialized model outperforms general-purpose multimodal

Reasoning and Math

- **Best capability:** Claude with extended thinking; OpenAI's GPT-5.4/5.5 Thinking (the line that absorbed the o-series); DeepSeek V4 with thinking mode — all competitive
- **Best cost:** DeepSeek V4 with thinking mode (much cheaper than equivalent GPT-5.x Thinking calls)
- **For formal mathematics:** DeepSeek-Prover-V2 is specialized and excellent for Lean-based formal work

Agent Workflows

- **Best capability:** Claude Opus and OpenAI's GPT-5.4/5.5 with native tool use are roughly tied
- **Best cost:** DeepSeek V4 Pro with thinking mode
- **For long-running production agents:** Claude has been particularly strong; DeepSeek catching up rapidly

The Pricing Landscape (Early 2026)

Approximate pricing per million tokens for the flagship variants of each provider:

Provider	Input (cache miss)	Input (cache hit)	Output

Provider	Input (cache miss)	Input (cache hit)	Output
OpenAI GPT-5.4/5.5	~$10	varies by impl	~$30
Anthropic Opus 4.7	~$15	~$1.50	~$75
Anthropic Sonnet 4.6	~$3	~$0.30	~$15
Google Gemini 3 Pro	~$2	~$0.50	~$12
Google Gemini 3.1 Pro	~$2	~$0.50	~$12
DeepSeek V4 Pro	~$0.43	~$0.0036	~$0.87
DeepSeek V4 Flash	~$0.14	~$0.0028	~$0.28

The pattern: DeepSeek V4 Flash is roughly **20–40× cheaper than premium OpenAI/Anthropic** on equivalent workloads. V4 Pro is roughly 10–15× cheaper. Gemini sits in the middle. DeepSeek defines most of the cost-per-quality frontier.

These numbers move. Check current pricing pages before making decisions based on exact figures. The order-of-magnitude relationship is the durable insight; the specific multipliers shift quarterly.

Operational Trade-offs Beyond Capability

A few operational dimensions that aren't about raw model capability but still matter substantially:

Reliability and Uptime

- **OpenAI:** Generally good uptime; high-profile outages happen but are rare
- **Anthropic:** Generally reliable; smaller scale than OpenAI means less battle-testing
- **Google:** Reliable; the underlying infrastructure is Google's, which is a high bar
- **DeepSeek:** Reliability has improved substantially through 2025; not at the level of the major Western providers but typically within acceptable range for production use

Rate Limits and Scaling

- **OpenAI/Anthropic:** Account-tier-based limits that scale with usage history; enterprise customers can get high limits
- **Google:** Similar pattern; quotas more visible in the GCP console
- **DeepSeek:** Limits exist but have generally been generous; enterprise customers can negotiate

Documentation and Developer Experience

- **OpenAI:** Best documentation, most third-party tutorials, most Stack Overflow answers
- **Anthropic:** Good documentation; smaller community than OpenAI
- **Google:** Documentation has improved substantially; complicated pricing/quota model
- **DeepSeek:** Docs adequate but less polished than competitors; community knowledge growing rapidly

Compliance and Enterprise Features

- **OpenAI:** Most mature enterprise features (SOC 2, HIPAA, BAA available); broad regulatory coverage
- **Anthropic:** Enterprise features available; positioning around safety appeals to regulated industries
- **Google:** Strong compliance posture for organizations already using Google Cloud
- **DeepSeek:** Less mature on enterprise compliance features; data jurisdiction concerns for Western enterprises

Geographic Considerations

- **OpenAI:** US-headquartered; varies by user region
- **Anthropic:** US-headquartered; some regions excluded
- **Google:** Global; multi-region deployment available
- **DeepSeek:** China-headquartered; data jurisdiction implications discussed in Chapter 31

A Multi-Provider Strategy That Uses Each for What It's Best At

A sample multi-provider architecture that pairs each provider with the workloads where it excels:

```
Workload Type                    →  Provider/Model
═══════════════════════════════════════════════════════
Routine chat / Q&A               →  DeepSeek V4 Flash
Customer support                 →  DeepSeek V4 Pro
Long-document analysis           →  Claude Opus 4.7
Complex coding / refactoring     →  Claude Opus 4.7
Routine code generation          →  DeepSeek V4 Pro
Code completion (low-latency)    →  DeepSeek V4 Flash (or self-hosted)
Multi-modal (vision)             →  GPT-5.x with vision or Gemini 3.1 Pro
Document OCR                     →  DeepSeek-OCR-2 (specialized)
Long-context (1M+ tokens)        →  Gemini 3.1 Pro
Agent workflows                  →  DeepSeek V4 Pro w/ thinking mode
Highest-stakes reasoning         →  Claude Opus 4.7
Embeddings                       →  OpenAI text-embedding-3 or local
```

This kind of routing isn't right for every team. Smaller teams often consolidate on one or two providers. Larger teams benefit from the complexity. The point isn't the specific assignment — it's that no single provider is the right answer for every workload.

How to Decide for Your Specific Use Case

A practical framework:

Step 1: Catalog your workloads. List the distinct AI use cases in your application. For each: what's the request shape? What quality matters? What latency is required? What's the volume?

Step 2: Identify capability requirements per workload. What's the minimum capability tier that's acceptable? Not "what would I love to use" — what's the actual minimum.

Step 3: Filter by operational requirements. Compliance, jurisdiction, latency, integration constraints. Eliminate providers that don't meet hard requirements for each workload.

Step 4: Compare cost across acceptable providers. For workloads where multiple providers meet the requirements, compare cost. The winner is usually clear.

Step 5: Build the abstraction layer. Whatever your provider mix, build the layer that lets you switch and route. Don't lock into a single provider's SDK shape.

Step 6: Re-evaluate quarterly. The provider landscape shifts. New models ship; pricing changes; capabilities evolve. The right answer six months from now may not be the right answer today.

For most teams, this exercise produces a multi-provider architecture with DeepSeek handling the largest share of traffic by volume and cost, and one or two premium providers handling the workloads where they genuinely outperform.

The Honest Conclusion

DeepSeek isn't categorically the best AI provider. It's also not categorically the cheapest substitute. It's a specific point on the capability-cost-trust frontier with a particular shape.

For most production workloads, that shape is favorable. For the hardest workloads, premium providers still win on capability. For the most regulated workloads, jurisdiction matters more than cost.

The right answer is almost always "DeepSeek plus something else." The something else depends on your specific work.

That closes Part VI. You now have a calibrated view of where DeepSeek fits in the broader landscape — cost and capability — and the architectural patterns for using it alongside other providers.

Part VII covers the specialized models — Prover-V2 for formal mathematics, Janus for multimodal vision, DeepSeek-OCR for document processing. If you have specific use cases that map to these specializations, Part VII is where the depth lives. Otherwise, you can skip ahead to Part VIII for the limitations and ethics treatment.

Part VII: Specialized Models

"Specialization beats generalization for narrow problems."

— Working principle

Chapter 27: DeepSeek-Prover-V2 (Formal Mathematics)

DeepSeek-Prover-V2 is a specialized model trained for one specific job: generating and verifying formal mathematical proofs within the Lean theorem prover framework. The audience is small but real — working mathematicians using formal proof assistants, formal verification engineers in software and hardware, and the growing population of computer scientists who treat mathematics formally.

For that audience, Prover-V2 is genuinely state-of-the-art among open-weight models. For everyone else, this chapter is a window into a niche application of large language models that hints at where AI in formal domains is heading.

What Formal Mathematics Means

Most mathematics is communicated informally — proofs in journal articles, textbook derivations, blackboard arguments. Other mathematicians read and judge the proofs, follow the reasoning, and accept (or challenge) the conclusions.

Formal mathematics is different. It expresses proofs in a fully rigorous logical language that can be machine-verified. Every step of a formal proof is checked by a computer program (a **proof assistant**) that confirms the logical correctness of each inference. No room for hidden assumptions, hand-waving, or intuitive leaps that don't actually hold up.

The major proof assistants in active use as of 2026:

- **Lean 4** — increasingly the dominant choice for new formal mathematics work; particularly strong in mathlib (the formalized mathematics library)
- **Coq** — long-established; strong in computer science applications
- **Isabelle/HOL** — long-established; strong in software verification
- **Agda** — dependently-typed; strong in programming language theory

Formalizing mathematics in any of these is laborious. A theorem that takes a paragraph to prove informally might take pages of formal Lean code. The promise of AI assistants like Prover-V2 is closing that gap — letting mathematicians work at the speed of informal proof while getting the rigor of formal verification.

What Prover-V2 Does

Prover-V2 is trained specifically on the corpus of Lean 4 proofs (especially mathlib) and produces Lean-syntax proofs as output. In practice, its capabilities:

Theorem-to-proof generation. Given a stated theorem in Lean syntax, Prover-V2 generates a candidate proof. Lean itself verifies the proof; if it checks, you have a confirmed formal proof.

Tactic-level guidance. When working interactively in Lean, Prover-V2 can suggest specific tactics for the current proof state — useful when you're stuck in the middle of a longer proof.

Proof completion. Given a partial proof, Prover-V2 fills in the remaining steps.

Lemma suggestion. Identifies which existing mathlib lemmas might apply to a current goal.

Translation from informal to formal. Take a sketched informal proof; Prover-V2 attempts to formalize it into Lean syntax. Quality varies — works well for proofs that closely follow standard patterns, struggles for proofs that require novel techniques.

The output isn't always correct. Prover-V2 (like all current AI proof assistants) generates candidate proofs that may or may not type-check. The workflow is: generate, verify, iterate.

How It Compares to Other Approaches

A few comparison points for context:

Vs. tactic-only systems. Older AI-for-Lean approaches used specialized neural networks trained to suggest individual tactics. Prover-V2 generates full proofs (or substantial portions), a different mode of operation. Both have value; full proof generation saves more time when it works.

Vs. general-purpose LLMs. GPT-5.5 and Claude can attempt Lean proofs but lack the focused training. They generate plausible-looking Lean syntax that often fails to type-check. Prover-V2's specialized training produces a much higher rate of valid proofs.

Vs. closed-source competitors. Some closed-source systems (academic projects, commercial offerings) compete with Prover-V2 on specific benchmarks. Prover-V2's open-weight release is its primary differentiator — researchers can build on it, fine-tune it, study its failures.

Using Prover-V2

The model is available on Hugging Face under the same MIT license as DeepSeek's other releases. Practical usage paths:

Local inference for individual mathematicians. Run Prover-V2 on a workstation GPU using vLLM or similar. Integrate with your Lean editor (typically VS Code with the Lean extension) via a custom plugin or LSP integration.

Hosted inference for teams. Deploy Prover-V2 on shared infrastructure for a research group. Multiple users hit the same deployment for proof assistance.

Batch processing. For large formalization projects (e.g., systematically working through a textbook), batch-generate candidate proofs and verify them with Lean. The workflow becomes "review and accept" rather than "write from scratch."

A Worked Example: Simple Theorem

A simple example to make this concrete. Suppose you want to prove in Lean 4 that if `n` is even, then `n + 1` is odd.

You'd state the theorem:

```
theorem even_succ_odd (n : ℕ) (h : Even n) : Odd (n + 1) := by
  sorry
```

The `sorry` is a placeholder — Lean accepts the theorem as stated but flags that the proof is missing.

Submitting this to Prover-V2 (with appropriate setup), you'd get back something like:

```
theorem even_succ_odd (n : ℕ) (h : Even n) : Odd (n + 1) := by
  rcases h with ⟨k, hk⟩
  exact ⟨k, by omega⟩
```

Lean type-checks this and confirms the proof. Done.

For trivial theorems like this, manual proof is similarly fast. The benefit of Prover-V2 scales with proof complexity — for proofs that would take a mathematician an afternoon to formalize, having an AI generate a candidate that's correct or close-to-correct is hours saved.

Use Cases Beyond Pure Mathematics

Formal proof techniques aren't confined to pure mathematics. A few adjacent applications:

Smart contract verification. Smart contracts (especially in DeFi) handle large sums of money; bugs are catastrophic. Formal verification using Lean or similar lets you prove contracts have specific safety properties. Prover-V2 helps here the same way it helps with pure math — generating candidate proofs of safety properties.

Critical software verification. Some software (avionics, medical devices, security-critical systems) is formally verified rather than just tested. The formal verification work resembles mathematical proof work; AI assistance applies similarly.

Cryptographic protocol verification. Proving cryptographic protocols correct under specific threat models is formal work that benefits from AI assistance.

Type theory and programming language research. Researchers proving properties of type systems or programming languages use Lean and similar tools heavily.

For these adjacent communities, Prover-V2 is a relevant tool even when "formal mathematics" sounds far from their domain.

Limitations to Know

Honest framing of where Prover-V2 falls short:

Hard for novel proofs. When the theorem requires a genuinely novel proof technique (something not well-represented in the training data of existing proofs), Prover-V2 struggles. It excels at "this proof should follow standard patterns"; it stumbles on "this proof requires a new idea."

Tactic syntax stability. Lean's tactic syntax has evolved across versions. Prover-V2 trained on a specific snapshot. Proofs it generates may use tactics that work in one Lean version but not another. Pin your Lean version when working with the model.

Computer algebra isn't its job. Prover-V2 is for formal proofs, not symbolic computation. Use a CAS (Mathematica, SageMath, SymPy) for the latter; combine the two when needed.

Mathlib coverage. The model's familiarity with mathlib lemmas reflects what was in mathlib at training time. Newly added lemmas may not be suggested. For the most current mathlib usage, augment with manual lemma search.

Verification is non-negotiable. Prover-V2 generates candidate proofs. Lean verifies them. Don't trust a proof until Lean has accepted it. This isn't optional — it's the entire point of using a proof assistant in the first place.

When to Use Prover-V2

The clear cases:

- You're a working mathematician using Lean for formalization
- You're a formal verification engineer for software, hardware, or smart contracts
- You're a researcher in formal methods, type theory, or related fields
- You're a student learning Lean and want AI assistance with exercises
- You're an educator preparing formal-methods curriculum

The cases where it's not the right tool:

- General mathematical reasoning (use V4 with thinking mode)

- Symbolic computation (use a CAS)
- Mathematical writing or exposition (use a general LLM with strong writing)
- Numerical computation (use code execution + appropriate libraries)

For the niche audience that uses formal proof tools, Prover-V2 is a real productivity tool. For everyone else, it's an interesting demonstration of what specialized models can do — and a hint at where the next decade of AI-augmented formal methods is heading.

What Comes Next

The next chapter covers a different specialized lineage — Janus, DeepSeek's multimodal vision models. The audience is broader than Prover-V2's; many applications benefit from visual understanding.

Chapter 28: Janus and Multimodal DeepSeek

Most of this book has covered DeepSeek's text-and-code work. The vision side of the lineup — the **Janus** family of multimodal models — is its own story. Janus models take images as input, produce both text and image outputs, and handle workloads where understanding what's in a picture is part of the task.

For teams building applications with visual understanding requirements, Janus is the relevant DeepSeek lineage. This chapter covers what Janus does, how it compares to the alternatives (GPT-5.x with vision, Gemini's vision capabilities, Claude's image handling), how to set it up, and the workloads where it's the right choice.

What Janus Is

Janus is the brand name for DeepSeek's multimodal model line. The lineup has evolved — earlier Janus models, Janus-Pro variants, ongoing iterations — but the core concept holds: models that handle text and images jointly, with both understanding (image-as-input) and generation (image-as-output) capabilities.

The capabilities, in practice:

Image understanding. Provide an image; the model describes it, answers questions about it, extracts information from it. Standard multimodal-LLM behavior.

Document understanding. PDFs with charts, photos of documents, screenshots. The model treats them as visual content that includes text.

OCR and text extraction. Reading text from images. Better than dedicated OCR for some content; we'll get to DeepSeek-OCR-2 — the dedicated specialist — in the next chapter.

Image generation. Given a text prompt, Janus produces an image. Meaningful, but not at the level of dedicated image generators (Midjourney, FLUX, OpenAI's DALL-E and Sora).

Multimodal reasoning. Reasoning that mixes visual and textual context — "what's wrong with this UI mockup based on the requirements I described."

Comparing Janus to the Vision Alternatives

Where Janus stands as of early 2026:

vs. GPT-5.x with vision — OpenAI's flagship multimodal models (GPT-5.4/5.5 with vision, building on the GPT-4o lineage) remain the consumer-facing benchmark. Janus is competitive on general image understanding but typically behind on the hardest visual reasoning tasks. For most consumer applications GPT-5.x still wins; for production back-end multimodal workloads where cost matters, Janus is increasingly competitive.

vs. Gemini 3/3.1 Pro — Gemini's multimodal capabilities are particularly strong for video understanding and long-context vision. For workloads where you're feeding many images or video content, Gemini often wins. For single-image tasks, the gap narrows and Janus's cost advantage may favor it.

vs. Claude (Sonnet 4.6, Opus 4.7) — Claude handles images well as input but doesn't generate them. For pure image understanding, Claude is often the highest-quality option, at a substantial cost premium over Janus.

vs. Specialized vision tools — For specific niches (OCR, document layout analysis, object detection), specialized tools — including DeepSeek's own OCR model, covered next chapter — often outperform any general multimodal LLM. Don't reach for a general vision LLM where a specialized model exists.

Image Understanding Patterns

For workloads where you're handing images to the model to understand and reason about:

Document Image Q&A

```
import base64
from openai import OpenAI

client = OpenAI(
    api_key=os.environ["DEEPSEEK_API_KEY"],
    base_url="https://api.deepseek.com/v1",
```

```
)

def encode_image(path):
    with open(path, "rb") as f:
        return base64.b64encode(f.read()).decode()

response = client.chat.completions.create(
    model="janus-pro",  # check current docs for exact model name
    messages=[
        {
            "role": "user",
            "content": [
                {"type": "text", "text": "What does this invoice say is the tota
l amount due?"},
                {"type": "image_url", "image_url": {"url": f"data:image/png;base
64,{encode_image('invoice.png')}"}},
            ],
        },
    ],
)
print(response.choices[0].message.content)
```

The pattern matches OpenAI vision API conventions — multipart user content with text and image components. For document images, OCR happens implicitly inside the multimodal model.

Chart and Diagram Understanding

```
response = client.chat.completions.create(
    model="janus-pro",
    messages=[
        {
            "role": "user",
            "content": [
                {"type": "text", "text": "Describe this chart. What's the trend?
 What's the most surprising data point?"},
                {"type": "image_url", "image_url": {"url": f"data:image/png;base
64,{encode_image('chart.png')}"}},
            ],
        },
    ],
)
```

Chart understanding is a common multimodal use case in business reporting. Janus handles standard chart types (bars, lines, pies, scatter) reasonably well; complex visualizations or unusual chart types may need clarifying prompts.

Visual UI/UX Review

```
response = client.chat.completions.create(
    model="janus-pro",
    messages=[
        {
            "role": "system",
            "content": "You are a UX expert. Review the provided screenshot for
usability issues, accessibility concerns, and visual clarity problems."
        },
        {
            "role": "user",
            "content": [
                {"type": "text", "text": "Review this screen."},
                {"type": "image_url", "image_url": {"url": f"data:image/png;base
64,{encode_image('ui-mockup.png')}"}},
            ],
        },
    ],
)
```

Having an AI review UI screenshots gives useful first-pass feedback in a development workflow. It's not a substitute for human UX review, but it's a useful starting point.

Screenshot-Based Debugging

```
response = client.chat.completions.create(
    model="janus-pro",
    messages=[
        {
            "role": "user",
            "content": [
                {"type": "text", "text": "Here's an error my user is seeing. Wha
t does the error say and what should I do about it?"},
                {"type": "image_url", "image_url": {"url": f"data:image/png;base
64,{encode_image('error-screenshot.png')}"}},
            ],
        },
```

```
    ],
)
```

In support workflows, user-submitted screenshots become far more useful when an AI can read and interpret them.

Multi-Image Conversations

Most multimodal applications involve multiple images per conversation — a sequence of photos, a comparison, a series of screenshots. The pattern is straightforward: include multiple images in a single user message, or spread them across multiple turns.

```
messages = [
    {
        "role": "user",
        "content": [
            {"type": "text", "text": "Compare these two product photos."},
            {"type": "image_url", "image_url": {"url": photo_1_url}},
            {"type": "image_url", "image_url": {"url": photo_2_url}},
        ],
    },
]
```

A few practical notes:

Image sizing matters. Very large images (multi-megabyte, high-resolution) take longer to process and cost more. For most use cases, downsizing to ~1024px on the long edge is a good default.

Image format. PNG, JPEG, WEBP, and GIF are typically supported. PDFs aren't usually multimodal-API material — extract pages as images first.

Limits on images per request. Each provider has a max-images-per-request limit. For Janus this varies by version; check current docs. To batch-process many images, spread them across multiple requests.

Image Generation

Janus can generate images from text prompts. The generation API is typically separate from chat completion:

```
response = client.images.generate(
    model="janus-pro",
    prompt="A serene mountain landscape at sunrise, watercolor style",
    n=1,
    size="1024x1024",
)
print(response.data[0].url)
```

For most users, dedicated image generators (Midjourney, FLUX, DALL-E) produce higher-quality outputs than Janus. Janus's image generation is more capability checkbox than primary use case for most workloads.

Where Janus image generation makes sense:

- **Fully open-source pipelines.** When you want everything in your stack to be open-weight, Janus fits.
- **Self-hosted image generation.** For privacy or compliance reasons, generating images locally rather than via a hosted API.
- **Integrated workflows.** When the same model handling text understanding also produces images, the integration is simpler than juggling multiple providers.

For most consumer-facing image generation needs, the dedicated tools win on quality.

Self-Hosting Janus

Janus models are released as open weights, like DeepSeek's text models. Self-hosting follows the same patterns as Chapter 21 — vLLM (or another inference engine that supports multimodal models) for the inference layer, BentoML or your own service framework for production deployment.

Hardware requirements vary by Janus variant. Smaller variants run on consumer GPUs; larger ones need workstation or datacenter hardware. Multimodal models tend to be more memory-intensive than text-only models of comparable parameter count because of the visual processing components.

For teams with vision workloads sensitive enough to require self-hosting, this is a practical option.

Production Patterns

For production multimodal applications using Janus:

Validate image inputs. User-provided images may be malformed, corrupt, or oddly sized. Preprocess (resize, normalize, format-convert) before sending to the API.

Handle the cost asymmetry. Image inputs are typically more expensive than text-only inputs. Cache where possible (some providers cache image embeddings); don't reprocess the same image repeatedly.

Mind the latency. Multimodal calls are slower than text-only calls. In interactive workflows, that affects UX. Stream responses where possible; show loading states; don't block the UI.

Give the model visual context. "What's in this image?" is vague. "What's the main subject of this product photo, and is it clearly visible against the background?" gets better results.

Specify output structure. As with text-only workloads, JSON mode on image inputs gives you structured outputs you can consume programmatically.

Limitations to Know

Janus, like all multimodal LLMs as of early 2026, has real limits:

Spatial reasoning. "What's to the left of the chair?" questions are inconsistent. Multimodal models often struggle with precise spatial relationships.

Counting. "How many people are in this photo?" produces approximate answers. For precise counts, use specialized object-detection models.

Detail-level questions. "What's the exact text in fine print at the bottom?" may fail for very small text. Pre-OCR with a dedicated tool may be necessary.

Generation precision. Image generation matches the prompt's general intent but may get specific details wrong (number of fingers, exact text, precise composition).

Bias. Like other multimodal models, Janus inherits visual biases from its training data. Outputs may reflect them.

When Janus Is the Right Tool

The clear cases:

- You need multimodal understanding (image input + text output) at DeepSeek's cost levels
- You need self-hosted multimodal capability and don't want to depend on closed APIs
- You're building applications where image understanding is one piece of a broader text-heavy workflow (a customer support bot that occasionally sees screenshots, a research tool that occasionally analyzes diagrams)
- You're already in DeepSeek's ecosystem, and adding multimodal is a natural extension

When to reach for alternatives:

- For the highest-quality consumer multimodal applications, GPT-5.x with vision or Gemini 3.1 Pro may justify their cost
- For dedicated OCR, use DeepSeek-OCR-2 (next chapter) — better at the specific task
- For dedicated image generation, use Midjourney, FLUX, or DALL-E
- For video understanding, Gemini's video capabilities currently lead

What Comes Next

The next chapter covers DeepSeek-OCR-2, the specialized model that often outperforms general-purpose multimodal models on the specific task of extracting text from documents. For teams building document-processing pipelines, OCR-specific tools matter.

Chapter 29: DeepSeek-OCR-2 and Specialized Pipelines

Most teams that need to extract text from documents end up using one of three things: a traditional OCR engine (Tesseract is the most common open-source option), a cloud OCR service (AWS Textract, Google Document AI), or a general-purpose multimodal LLM (GPT-5.x with vision, Janus, Gemini's vision). Each has trade-offs.

DeepSeek-OCR-2 (released January 27, 2026, building on the original DeepSeek-OCR) sits in a useful niche: a 3B-parameter vision-language model trained specifically for high-accuracy document extraction, with capabilities traditional OCR engines lack (understanding document structure, handling complex layouts, managing handwriting and unusual fonts) at quality competitive with or better than general-purpose multimodal LLMs for this specific task. The DeepEncoder V2 architecture introduces a "causal flow" approach to visual encoding — the model dynamically rearranges image segments based on document meaning rather than scanning fixed top-left to bottom-right.

This chapter covers what DeepSeek-OCR-2 does, how it compares to the alternatives, and the practical patterns for building document-processing pipelines around it.

What DeepSeek-OCR-2 Does

The single-line description: it converts document images into structured text, preserving layout and handling content that traditional OCR struggles with.

The capabilities, in detail:

Multi-column documents. Academic papers, magazine layouts, multi-column reports. Traditional OCR often gets the reading order wrong (mixing columns); DeepSeek-OCR-2 understands the document structure.

Tables. Extracts tabular data with rows and columns intact. Traditional OCR tends to flatten tables into linearized text; DeepSeek-OCR-2 preserves structure.

Mixed content. Documents with headers, footers, sidebars, captioned images. The model handles each layout component separately rather than concatenating them naively.

Handwriting. Reasonably clear handwriting. Quality scales with clarity — neat handwriting works well, scrawled notes less so, but still better than most traditional OCR.

Mathematical notation. Equations, formulas, scientific notation. A meaningful improvement over general OCR for technical documents.

Foreign scripts. Many non-Latin scripts (Chinese, Japanese, Arabic, Cyrillic, and others). Coverage varies; check for your specific language needs.

Form structure. Forms with labeled fields, checkboxes, signatures. The model can extract field labels and values as paired data.

Output is typically structured text (Markdown, JSON, or plain text depending on configuration) that preserves the document's logical structure rather than just its visual rendering.

Comparing OCR Approaches

Where DeepSeek-OCR-2 fits in the landscape:

vs. Tesseract — Tesseract is the long-standing open-source OCR engine. Strong at clean text on simple layouts; weak at complex layouts, mixed content, and anything with unusual fonts or handwriting. DeepSeek-OCR-2 substantially outperforms it on most real-world documents.

vs. AWS Textract/Google Document AI — Cloud OCR services offer specialized features (form extraction, table extraction, key-value detection) at moderate per-page cost. DeepSeek-OCR-2 is competitive on quality, generally cheaper at scale, and has the open-weights option for self-hosting. Cloud services have more mature enterprise features (compliance certifications, dedicated support); DeepSeek is leaner.

vs. Janus/general multimodal LLMs — General multimodal models can do OCR but aren't optimized for it. They sometimes hallucinate text that isn't there, miss content, or get layout wrong. DeepSeek-OCR-2's specialized training produces more reliable extraction for the specific task.

vs. Premium multimodal (GPT-5.x with vision, Claude with vision) – Premium models can do high-quality OCR on well-formed documents. DeepSeek-OCR-2 is competitive at substantially lower cost. For very challenging documents, premium models may still edge ahead; for routine document processing at scale, DeepSeek-OCR-2 is the cost-effective choice.

For most document-processing pipelines, the choice is between DeepSeek-OCR-2 (cost-effective, open-weight), cloud OCR (managed service with enterprise features), or premium multimodal LLMs (highest quality at highest cost). The right choice depends on volume, document complexity, and operational requirements.

Practical Use

DeepSeek-OCR-2 can be used via API (similar pattern to other DeepSeek models) or self-hosted using the open weights.

API Usage

```
import base64
from openai import OpenAI

client = OpenAI(
    api_key=os.environ["DEEPSEEK_API_KEY"],
    base_url="https://api.deepseek.com/v1",
)

def encode_image(path):
    with open(path, "rb") as f:
        return base64.b64encode(f.read()).decode()

response = client.chat.completions.create(
    model="deepseek-ocr",  # check current docs for exact model name
    messages=[
        {
            "role": "user",
            "content": [
                {"type": "text", "text": "Extract all text from this document. P
reserve the layout structure."},
                {"type": "image_url", "image_url": {"url": f"data:image/png;base
64,{encode_image('document.png')}"}},
            ],
```

```
        },
    ],
)
print(response.choices[0].message.content)
```

The exact API shape depends on the deployment — check current documentation for parameter specifics.

Self-Hosted Usage

For high-volume document processing, self-hosted DeepSeek-OCR-2 can be substantially cheaper than API calls. Setup follows the patterns from Chapter 21 — vLLM or a similar inference engine with the OCR model loaded.

For batch processing thousands or millions of documents, self-hosted is typically the right architecture. For occasional or moderate volume, the API is simpler.

Document-Processing Pipelines

OCR is rarely the whole pipeline — it's typically the first step, with downstream processing extracting structured data, validating results, and feeding into business systems.

A Standard Document Pipeline

```
Document (PDF or image)
         |
         ▼
┌─────────────────┐
│  Preprocessing  │  ← deskew, denoise, normalize size
└─────────────────┘
         |
         ▼
┌─────────────────┐
│ DeepSeek-OCR-2  │  ← extract text + structure
└─────────────────┘
         |
         ▼
┌─────────────────┐
│  Structured     │  ← LLM call to extract specific fields
│  Extraction     │    into JSON
└─────────────────┘
         |
```

```
          ▼
┌─────────────────┐
│   Validation    │  ← schema check, business rule validation
└────────┬────────┘
         │
         ▼
┌─────────────────┐
│     Storage     │  ← database, downstream system
└─────────────────┘
```

A few key practices for each stage:

Preprocessing. Many real-world documents need cleanup before OCR — deskewing, denoising, contrast adjustment, splitting multi-page PDFs into individual pages. `pdf2image`, OpenCV, and similar libraries handle this in Python.

OCR (DeepSeek-OCR-2). Pass the preprocessed image; get back structured text. For multi-page documents, batch the pages.

Structured extraction. OCR output is text with layout — usually still needing further processing to extract the specific fields you care about. Use JSON-mode LLM calls (Chapter 14) to pull out structured data.

Validation. Schema-check the extracted JSON, validate business rules (does this date make sense, is this dollar amount in a reasonable range), flag low-confidence extractions for human review.

Storage. Persist to your downstream systems — databases, document management, analytics warehouses.

A Specific Example: Receipts Pipeline

Putting it together, a receipts-to-spreadsheet pipeline:

```
import json
from pathlib import Path
from openai import OpenAI

client = OpenAI(
    api_key=os.environ["DEEPSEEK_API_KEY"],
    base_url="https://api.deepseek.com/v1",
)

def ocr_receipt(image_path):
```

```
    with open(image_path, "rb") as f:
        encoded = base64.b64encode(f.read()).decode()
    response = client.chat.completions.create(
        model="deepseek-ocr",
        messages=[
            {
                "role": "user",
                "content": [
                    {"type": "text", "text": "Extract all text from this receipt
 with layout preserved."},
                    {"type": "image_url", "image_url": {"url": f"data:image/jpeg;
base64,{encoded}"}},
                ],
            },
        ],
    )
    return response.choices[0].message.content

def extract_fields(ocr_text):
    response = client.chat.completions.create(
        model="deepseek-v4-flash",
        messages=[
            {
                "role": "system",
                "content": """Extract receipt fields from the OCR text. Respond
as JSON:
                {
                  "merchant": "<store name>",
                  "date": "<ISO 8601 date>",
                  "total_cents": <integer cents>,
                  "items": [{"description": "...", "price_cents": <integer>},
 ...],
                  "confidence": <0.0 to 1.0>
                }
                Use null for missing fields.""",
            },
            {"role": "user", "content": ocr_text},
        ],
        response_format={"type": "json_object"},
    )
    return json.loads(response.choices[0].message.content)

def process_receipt(image_path):
    ocr_text = ocr_receipt(image_path)
    fields = extract_fields(ocr_text)
```

```
    return fields

# Process a directory of receipts
results = []
for img in Path("receipts/").glob("*.jpg"):
    try:
        receipt = process_receipt(img)
        receipt["source_file"] = img.name
        results.append(receipt)
    except Exception as e:
        print(f"Failed to process {img}: {e}")

# Save as CSV or load into your accounting system
import csv
with open("receipts.csv", "w") as f:
    writer = csv.DictWriter(f, fieldnames=["source_file", "merchant", "date", "t
otal_cents", "confidence"])
    writer.writeheader()
    for r in results:
        writer.writerow({k: r.get(k) for k in writer.fieldnames})
```

A complete receipts-processing pipeline in about 50 lines. Cost per receipt runs in the cents, and throughput is limited by API latency rather than human attention.

When DeepSeek-OCR-2 Isn't the Right Choice

A few scenarios where alternatives make more sense:

Very simple text on clean backgrounds. Tesseract is faster and free for cases it handles well — printed text on white backgrounds with simple layouts. For these, DeepSeek-OCR-2 is overkill.

Real-time/low-latency requirements. OCR via API has network round-trip latency. For workloads requiring sub-second extraction (live document-scanning UIs), self-hosted or edge OCR may be necessary.

Highly specialized document types. Some document types have dedicated extractors (invoice processors, medical-records tools). For workflows entirely within these niches, the specialized tools may outperform general OCR.

Compliance-restricted environments. If sending document images to any external API isn't acceptable, self-hosting is the only option. DeepSeek-OCR-2's open weights enable that; cloud OCR services don't.

Highest-quality requirements with cost flexibility. Where every error matters and budget is unconstrained, premium multimodal LLMs may have a slight edge over DeepSeek-OCR-2. For most production workloads, the cost-quality trade-off favors DeepSeek-OCR-2.

Production Considerations

For document-processing pipelines running at scale, a few practices matter:

Preprocessing pipeline. A robust preprocessing step (deskew, denoise, split, normalize) dramatically improves OCR accuracy. Don't skip it.

Quality monitoring. Track confidence scores, validation failure rates, and downstream review rates. Sudden drops indicate either upstream document quality changes or model behavior shifts.

Human-in-the-loop for low-confidence cases. Build a review interface where humans validate extractions flagged as uncertain. The pipeline becomes a productivity tool for human reviewers rather than a replacement for them.

Audit trail. For regulated workloads (financial documents, medical records, legal documents), preserve the original document, the OCR output, and the extracted fields. Audit needs vary by industry; design for the strictest requirement.

Cost monitoring. Per-document cost is small but compounds across volume. Track it; investigate spikes.

What Comes Next

That closes Part VII. You now have a sense of where DeepSeek's specialized models fit — Prover-V2 for the niche of formal mathematics, Janus for multimodal text-and-image work, DeepSeek-OCR-2 for document processing.

For most users, only one or two will be relevant. For the audiences they fit, they're the right tool — better than general-purpose models, cheaper than cloud-service alternatives, with the open-weights option that distinguishes the DeepSeek lineage.

Part VIII is the section other DeepSeek coverage skips: an honest treatment of what the models get wrong, the China question and what it means for users, and the ethics of open-weight AI. If you're going to use DeepSeek for serious work, Part VIII is the part you can't afford to skim.

Part VIII: Safety, Ethics, and Limitations

"The honest treatment is the load-bearing one."

— A working principle

Chapter 30: What DeepSeek Gets Wrong

This is the chapter every honest book about an AI tool needs and most don't include. DeepSeek is a remarkable piece of engineering. It's also wrong, sometimes, in ways that matter. Knowing the failure modes is the difference between using the tool well and getting burned by it.

Most of this book has built the cost-and-capability case for DeepSeek. This chapter pulls in the opposite direction: a calibrated view of where it falls short, the failure modes that show up most often, and the mitigations that protect you from them.

Hallucinations and Confident Errors

Like every large language model, DeepSeek hallucinates — generates content that sounds plausible but isn't true. This shows up most often as:

Made-up facts, especially specifics. Dates, numbers, statistics, names of people, names of companies. The model knows roughly what kind of fact to produce; it sometimes produces one that doesn't actually exist.

Invented citations. Papers that don't exist, cited authors who never wrote on the topic, journal references that don't check out. Particularly dangerous because citations look authoritative.

Plausible-but-wrong code. Code that uses APIs that don't exist (hallucinated function names, made-up library methods), code that compiles but has subtle correctness bugs, code that mishandles edge cases.

Confidently stated misinformation. "X is true because Y" framings that make wrong answers feel reliable. Confidence isn't a signal of accuracy.

The mitigations:

Verify load-bearing claims. Anything you'll quote, cite, base a decision on, or stand behind in writing — check it against authoritative sources.

Use thinking mode for high-stakes calls. The reasoning trace makes errors more visible than direct generation does. You can catch hallucinations in the reasoning that direct outputs would hide.

Pair the model with verification tools. For code, run the code. For math, compute exactly. For citations, look them up. Don't trust the model alone where ground truth matters.

Set the model's expectations. "Only state things you're certain about; if uncertain, say so" prompts produce more measured outputs than asking for confident answers.

Know your domain. The model hallucinates more in narrow technical areas than in broad familiar ones. In a niche, dial your skepticism up.

Subtle Domain Errors

Beyond outright hallucination, DeepSeek (like all general-purpose LLMs) makes subtle errors in specialized domains. The pattern: content that sounds right to a non-specialist but contains errors a specialist would catch.

Most visible in:

- **Specialized professional fields.** Medicine, law, finance. The model produces plausibly correct but technically wrong content often enough that any application in these domains needs domain-expert review.
- **Cutting-edge research areas.** Where training data is sparse, model knowledge is sparse. Recent developments may be missed or distorted.
- **Specialized programming languages and frameworks.** Less popular languages, niche frameworks, internal company tools. The model fills in patterns from popular languages, often wrongly.
- **Cultural specifics.** Outside dominant Western contexts, model knowledge thins out. Cultural references, regional practices, local norms — the model approximates from limited data.

The mitigations are organizational rather than technical:

- **Domain experts review outputs in their domains.** No amount of prompting replaces the expert check.

- **Cross-reference with authoritative sources.** Use the model as a starting point; verify with primary documentation.
- **Be humble about your ability to spot errors in domains you don't know well.** The Dunning-Kruger risk with AI tools is real.

Long-Context Decay

DeepSeek's context window is large (1M tokens for V4). Quality across that range isn't uniform. The patterns:

Information buried in the middle of long contexts gets less attention. "Lost in the middle" effects, well-documented across LLMs, apply to DeepSeek too. Important content in the middle of a long input may be effectively missed.

Cross-references across long distances are unreliable. "Compare what the document said in section 3 to what it says in section 47" — for long documents, the model may not hold the section-3 details well enough to compare.

Coherence drift in very long generations. When generating 15K+ tokens, the model can drift from its earlier framing or contradict itself.

Mitigations:

Put critical information at the start or end. Don't bury the lede.

Ask cross-reference questions in chunks. Rather than "compare section 3 to section 47," ask about each separately and have the model synthesize.

For very long sources, split. Multiple targeted queries against sections often outperform one query against the whole document.

For very long generations, structure them. Ask for outlines first, then expand sections. Structured long-form beats long-form that tries to flow.

Reasoning Trace Failures

Even with thinking mode on, the reasoning trace isn't always reliable. A few patterns:

Reasoning wrong, answer right. The model sometimes reaches a correct answer through bad reasoning. Reading the trace can mislead you if you assume good answer = good reasoning.

Reasoning right, answer wrong. The opposite — careful reasoning that arrives at a wrong final answer due to a last-step error. Verify the answer separately even when the reasoning looks sound.

Performative reasoning. Sometimes the trace looks like reasoning but doesn't actually constrain the answer. The model produces plausible reasoning text and then produces the answer it would have produced anyway. Hard to detect from the outside.

Reasoning consumed without payoff. Some workloads pay for thinking-mode tokens without benefiting from the reasoning. If your task is simple enough that direct generation produces good answers, the reasoning trace is just cost.

Mitigations:

Spot-check the relationship between reasoning and answer. For high-stakes calls, read both critically. If the reasoning doesn't justify the answer, the answer is suspect even if both look reasonable on their own.

Use thinking mode where it's been validated. Test on representative samples before committing to thinking-mode-everywhere. Some workloads benefit; others don't.

Tool-Calling Failures

Agent workflows (Chapter 18) introduce additional failure modes:

Wrong tool selection. The model picks a tool that doesn't fit the request.

Wrong arguments. Right tool, mistyped or invalid arguments.

Loops. The model keeps calling tools without ever producing a final answer.

Hallucinated tool results. The model proceeds as if it had received a result it never got.

Brittle to tool errors. A tool returning an error sometimes produces nonsense rather than recovery.

Mitigations covered in Chapter 13 and Chapter 18: clearer tool descriptions, fewer tools, structured error returns, hard step limits, cost circuits.

Caching Surprises

For production systems relying on prompt caching (Chapter 12), a few caching-specific failure modes:

Cache invalidation from small prompt edits. A casual tweak to your system prompt destroys the entire cache. Costs spike unexpectedly.

Cache misses from variable content in the prefix. Timestamps, request IDs, or per-call variables in the prefix bust caching for every request.

Cache hit rate degradation over time. As your conversation history grows or your prompt structure evolves, hit rates can drop. Monitoring is the only way to catch this early.

Mitigations: prompt-change discipline (test cache impact before deploying), structural prompt design (variable content goes last), continuous monitoring of cache hit rates.

Censorship and Refusals

A failure mode specific to DeepSeek's hosted services: refusals on politically sensitive topics. Chapter 31 covers it in detail, but it's worth flagging here as a real source of behavior you might not expect:

- Questions about Tiananmen Square, the 1989 protests, related events
- Questions about Xinjiang, particularly Uyghur-related topics
- Questions about Taiwan's political status
- Questions about Hong Kong's political situation
- Questions about specific Chinese political figures
- Questions critical of the Chinese government

For these, DeepSeek's hosted API typically returns refusals or scripted responses rather than substantive engagement. Self-hosted DeepSeek (using the open weights) is less constrained but still reflects training-data biases.

For applications that might encounter these topics, plan accordingly — route those queries to a different provider, or design the application to handle refusals gracefully.

Multilingual Limits

DeepSeek handles English well. Other languages vary:

- **Major languages** (Spanish, French, German, Mandarin, Japanese, Portuguese): generally good quality
- **Less common languages**: variable; sometimes adequate, sometimes notably weaker
- **Low-resource languages**: capability often substantially below English

Evaluate empirically in each language you care about. Don't assume English-level quality elsewhere.

Cost Surprises

Production cost can deviate from forecasts in characteristic ways:

Output tokens grow unexpectedly. Prompt drift toward verbosity, retries that double output, unbounded responses.

Cache hit rates drop without obvious cause. Subtle prompt changes, conversation pattern shifts.

Thinking mode triggered more than expected. A new feature that defaults to thinking mode, or a routing heuristic firing more often than tested.

Agent loops running longer. Changes to tool behavior or model versions affecting how quickly agents reach completion.

Mitigations: cost monitoring per route, feature, and user; alerts on deviations from baseline; regular cost review with attribution.

Operational Failure Modes

Beyond the model itself, the operational reality of using DeepSeek's API:

Rate limit surprises. Bursts that exceed limits cause request failures. Unhandled, they break user experiences.

API outages. All cloud APIs have downtime. Build with retry, fallback, and graceful degradation.

Region availability. DeepSeek's API may not be available or performant from all regions. Test from your user locations.

Documentation lag. New features sometimes ship before documentation catches up. Watch the changelog and community discussions.

Behavior changes between model versions. Upgrading from V3.2 to V4 will change behavior even when you don't expect it to. Test before migration.

Honest Failure-Mode Summary

A composite picture of how DeepSeek fails in real-world use:

Most failures are silent. The model produces something that sounds right but isn't. Without verification, the failure propagates.

Failures concentrate in specific patterns. Specifics, citations, narrow domains, long contexts. Knowing the patterns helps you target your verification effort.

Failures are usually catchable. Verification, domain review, testing, monitoring — these aren't optional rituals; they're how you catch the failures before they cause real damage.

Failures aren't unique to DeepSeek. Most failure modes apply to all current LLMs. The mitigations transfer too.

What "Failure Mode Awareness" Looks Like in Practice

For users serious about deploying DeepSeek, the practical implications:

For developers: verification loops in production code. Don't trust generated code without running it. Don't trust extracted data without schema validation. Build the safety nets.

For analysts and researchers: cross-check against authoritative sources. The model is your starting point; primary sources are your ground truth.

For knowledge workers: read the citations. Click through. Confirm specifics. The discipline from Book 11 (NotebookLM's citation-verification chapter) applies here too.

For organizations: observability. Track quality metrics. Alert on degradation. Know your failure modes well enough to catch them at the system level rather than hearing about them from customers.

For everyone: calibrated trust. DeepSeek is genuinely useful across a wide range of work. It's not a substitute for the verification, expertise, and judgment that make work trustworthy.

Chapter 31: The China Question

Most coverage of DeepSeek skirts this topic. This chapter doesn't.

DeepSeek is a Chinese AI company. The hosted API runs on infrastructure in China. The training data reflects what the Chinese government considers acceptable to train on. The behavior of the hosted models reflects the political constraints of operating in China. None of this is incidental, and all of it has implications for users.

This chapter is the honest treatment: what data jurisdiction means in practice, what content the hosted services refuse, what self-hosting changes, the geopolitical context shaping the AI industry in ways DeepSeek is at the center of, and a framework for deciding whether to use DeepSeek for your specific work.

It isn't pro-DeepSeek or anti-DeepSeek. It's pro-clarity. Both the legitimate concerns and the legitimate value need honest acknowledgment if users are going to make informed decisions.

What "Chinese Provider" Means in Practice

DeepSeek (Hangzhou DeepSeek Artificial Intelligence Co.) is a Chinese company headquartered in Hangzhou, founded in 2023 by Liang Wenfeng. It operates under Chinese law, as do all Chinese tech companies.

The implications, in concrete terms:

Data residency. When you use DeepSeek's hosted API, your data flows to and is processed on infrastructure in China. The specific data centers and geographic locations within China may not be publicly documented.

Chinese law applies. Chinese cybersecurity law, data protection law, and national security law apply to data processed within China. That includes provisions requiring Chinese companies to cooperate with government data requests in ways that go beyond what Western companies typically face.

No jurisdiction-shifting. Unlike some cloud providers that offer data residency in specific regions (US-only, EU-only), DeepSeek's hosted API doesn't typically let you constrain processing to a specific non-Chinese region.

Standard commercial terms. DeepSeek operates under standard-looking commercial terms (rate limits, SLAs, payment terms). The legal substrate underneath those terms differs from Western providers in ways your legal team should understand.

For most casual users, none of this matters in practice. For users handling sensitive data, organizational data, or regulated workloads, it matters substantially.

Content Censorship

DeepSeek's hosted services (chat.deepseek.com and the hosted API) refuse or deflect on certain topics. The pattern is consistent and worth understanding before deploying:

Topics that produce refusals or scripted non-answers:

- **Tiananmen Square 1989 protests** — treated as off-limits. Direct questions get refusals. Indirect approaches sometimes get partial answers, but typically with sanitized framing.
- **Xinjiang and Uyghur-related topics** — refusals or PRC-aligned framings on questions about detention camps, cultural suppression, and related topics.
- **Taiwan's political status** — questions framed in ways that imply Taiwan is independent typically get refusals or restated into PRC-aligned framings.
- **Hong Kong's political situation** — similar pattern, especially for topics related to the 2019–2020 protests and the National Security Law.
- **Specific Chinese political figures** — particularly Xi Jinping. Critical commentary, comparisons to historical figures considered sensitive, and certain types of analysis produce refusals.
- **Falun Gong and other restricted religious or political movements** — refusals on topics treated as sensitive in mainland China.
- **Sensitive historical events** — Cultural Revolution criticism (varies in directness), Great Leap Forward death-toll discussions, and similar.

Topics that produce normal responses:

- General political analysis of non-Chinese topics
- Critical commentary on Western governments and politicians
- Analysis of geopolitical conflicts that don't directly implicate China's interests
- Most everyday questions completely unrelated to the above

The pattern isn't subtle. Test your specific use cases if you might hit sensitive topics. The refusals are consistent enough that you can plan around them.

Why the Censorship Exists

Honest framing: DeepSeek operates in China and is subject to Chinese regulations on AI products. Chinese law (including the Generative AI Service Management Provisions effective August 2023) requires AI services operating in China to:

- Train on data that aligns with "core socialist values"
- Refuse generation of content that "subverts state power" or "undermines national unity"
- Implement content filtering for politically sensitive topics

This isn't unique to DeepSeek. Every Chinese AI company operates under the same constraints. The behavior you see in DeepSeek's hosted services isn't a bug or an oversight — it's a condition of operating an AI product legally in China.

For users, this means the censorship isn't going to be fixed in a future update. It's structural. As long as DeepSeek operates in China and serves a Chinese audience, the constraints will apply.

Self-Hosting Changes the Picture (Mostly)

For users who self-host DeepSeek using the open weights, several of the concerns above shift:

Data jurisdiction. Your data doesn't leave your infrastructure. Whatever residency rules apply to your own systems apply to your DeepSeek deployment.

Censorship. The open weights are less aggressively constrained than the hosted services. Self-hosted DeepSeek will engage with topics the hosted API refuses.

Training-data-level biases remain (the model's "default position" on sensitive topics still reflects training), but the explicit refusal layer is largely absent.

Cooperation with Chinese authorities. Self-hosted models can't be queried for user data by Chinese authorities — the data isn't in their jurisdiction.

What doesn't change with self-hosting:

Training-data biases. The model was trained on data filtered for compatibility with PRC norms. Its default framings on certain topics will reflect that training. Fine-tuning can adjust this but doesn't eliminate it.

Geopolitical optics. "We use a Chinese-developed AI model" remains true even when self-hosted. Some customers, partners, and stakeholders will have opinions about that regardless of where the inference runs.

Provenance. The intellectual contribution behind the model came from a Chinese research lab. Whether that's relevant depends on your values and concerns.

For organizations whose primary concern is data jurisdiction, self-hosting addresses the issue cleanly. For organizations whose concerns are broader, self-hosting helps but doesn't eliminate them.

The Geopolitical Context

DeepSeek emerged in the middle of a multi-year period of AI-related US-China competition that's still unfolding. Context worth understanding:

Export controls. The US has imposed increasing restrictions on chip exports to China, particularly the most advanced AI training chips (NVIDIA's H100, H200, B200, and successors). The restrictions are intended to slow Chinese AI capability development.

Open-weights tension. DeepSeek's open-weight releases sit awkwardly in this context. The weights are freely downloadable from the US and elsewhere — circumventing export controls in spirit if not in letter. US policy responses remain uncertain.

Technology transfer concerns. US national security and economic policy has expressed concern that Chinese AI capabilities might benefit from access to US chip technology, research, and engineering talent. The open-weights model partially inverts this — Chinese AI capability is being shared globally, including back to US users.

Industry response. Some US AI companies have advocated for restrictions on the use of Chinese AI models in US infrastructure. As of early 2026, no major formal restrictions are in place, but the policy environment is fluid.

Reciprocity. US AI services (OpenAI, Anthropic, Google) are largely unavailable in China. Chinese AI services (DeepSeek included) are largely available globally. The asymmetry creates complicated dynamics.

For US-based users, the practical implications: using DeepSeek today is legal and unrestricted. Future regulatory actions are possible but not imminent. Track the policy environment if you're building production systems with substantial DeepSeek dependence.

A Framework for Deciding

How to think about whether DeepSeek is appropriate for your specific use case:

Step 1: Identify the data category.

Categories that warrant extra scrutiny: - Personal data of identifiable individuals (subject to GDPR, CCPA, or similar) - Health information (HIPAA-covered in the US) - Financial information (subject to financial sector regulations) - Legal materials (subject to attorney-client privilege) - Internal organizational confidential data - Government or defense-related information - Trade secrets and intellectual property

For these categories, the threshold for using DeepSeek's hosted API is high. Self-hosting may be acceptable; hosted API generally isn't.

Step 2: Identify the content category.

If your use case is likely to encounter: - Politically sensitive topics (especially China-related) - Critical commentary that DeepSeek's hosted services would refuse - Translation between languages where Chinese-trained model behavior matters

Plan for the censorship behavior. Either route those queries elsewhere or design for graceful handling.

Step 3: Identify the customer/stakeholder context.

If your customers, partners, regulators, or board would have concerns about Chinese AI provider use: - US government contracts often have specific restrictions - Some enterprise customers ask about AI provider sourcing - Some partnerships have provider-restriction clauses - Some boards have opinions at the oversight level

Know your context. The "what AI provider should we use" decision sometimes has political dimensions beyond the technical ones.

Step 4: Apply the right architecture.

Based on the above: - **Most concerns**: don't use DeepSeek; use Western providers instead - **Some concerns**: use DeepSeek for low-risk workloads; route sensitive ones to Western providers - **Few concerns**: use DeepSeek broadly with appropriate cost-routing patterns

The right architecture depends on your specific constraints. There's no universal answer.

Honest Use Cases by Risk Profile

A quick map of where different use cases land:

Low-risk uses (DeepSeek typically fine): - Coding assistance for non-sensitive code - General educational content generation - Personal productivity assistance - Open-source project work - Public-data research and analysis - Marketing content for general audiences

Medium-risk uses (consider carefully): - Customer-facing applications where customers might object to Chinese provider use - Content generation for sensitive industries (finance, healthcare) - Internal tooling for organizations with mixed-jurisdiction data - Workloads that occasionally touch sensitive topics

High-risk uses (DeepSeek generally inappropriate): - Government contracts with provider restrictions - Healthcare with HIPAA-protected information - Financial services with regulated data - Legal work with privileged communications - Sensitive personal data of identifiable individuals - National security or defense applications

These are guidelines, not rules. Your situation may differ. The framework is: consider the data, the content, the stakeholders, and pick architectures that address the legitimate concerns.

What This Doesn't Mean

A few things this chapter is not arguing:

Not arguing DeepSeek is bad. The model is a remarkable engineering achievement. Its cost-economics impact on the AI industry is genuine and broadly positive for users. The open-weights movement DeepSeek represents has substantial merit.

Not arguing all Chinese AI is suspect. The concerns are specific (data jurisdiction, content censorship, geopolitical context) — not categorical.

Not arguing Western AI providers are always better. They have their own issues (different jurisdictional concerns, different content policies, different commercial dependencies). The comparison is workload-specific, not categorical.

Not arguing the politics will stay where they are. The geopolitical context is fluid. What's true today may shift. Track developments.

Not arguing you should use or not use DeepSeek. The right answer depends on your situation. The chapter gives you the framework; the decision is yours.

A Personal Synthesis

For most readers of this book, DeepSeek is a tool that's genuinely useful for substantial parts of their work, with constraints that matter for other parts. The honest path is to:

- Use DeepSeek for the workloads where it fits
- Use other providers for the workloads where it doesn't
- Maintain awareness of the broader context as it evolves
- Be transparent with stakeholders about provider choices

This isn't a chapter about whether DeepSeek deserves to exist or whether you should boycott it. It's a chapter about using powerful tools responsibly in a complicated geopolitical context. The complications are real. The tool is also real. Both deserve clear-eyed treatment.

The next chapter takes on the broader ethical questions about open-weight AI — what changes when frontier capability is freely downloadable, what the misuse risks are, and how to think about the future of the open-weights movement DeepSeek is at the center of.

Chapter 32: The Ethics of Open-Weight Models

This is the last full chapter of the book. The previous chapter covered the DeepSeek-specific political dimensions; this chapter goes broader, into the ethical questions raised by the open-weights movement DeepSeek is at the center of.

The questions here don't have settled answers. The open-weights debate is one of the most contested topics in AI policy and ethics in 2026, with serious people on multiple sides taking serious positions. The chapter lays out the questions honestly rather than pretending consensus where none exists.

What "Open-Weight" Means and Why It Matters Ethically

A quick recap. **Open-weight** AI models have their trained parameters (the weights themselves) released publicly, typically under permissive licenses. Anyone can download them, run them locally, modify them, and build on them.

Closed-weight models — most major commercial AI services — keep the weights proprietary. Users access the models only through APIs the provider operates. The provider retains control over capability, behavior, and access.

The ethical implications differ substantially. Closed-weight models centralize control: the provider can update the model, modify its behavior, restrict its use, and retire it. Open-weight models distribute control: once weights are released, no one can fully retract them.

For frontier-capability AI, this distinction carries stakes that go beyond ordinary software-licensing questions.

The Case For Open Weights

The proponents of open-weight AI advance several arguments:

Democratization of access. Open weights let researchers, students, small companies, and individuals access frontier-capability AI without paying ongoing

API fees. The ecosystem of derivative work, fine-tuning, education, and research that emerges around open weights can't develop around closed ones.

Reproducibility of research. Scientific research that uses AI models is reproducible only if other researchers can access the same models. Open-weight models support reproducible research; closed-weight models inherently don't.

Reduced concentration risk. When frontier AI capability is concentrated in a few companies, those companies have outsized influence over what AI gets used for and how. Open-weight models distribute capability more broadly.

Auditability and safety research. Open weights enable independent safety research — probing for failure modes, testing for biases, identifying capabilities the original developers might have missed. Closed weights make this kind of research impossible.

Resilience. When models are open-weight, no single company's decision (deprecating a model, raising prices, changing terms of service) can disrupt the work that depends on them. The community can keep using and maintaining released models indefinitely.

Sovereignty. Countries, organizations, and individuals can operate AI capability without depending on infrastructure controlled by entities they don't fully trust. For some users, that's a fundamental requirement.

These arguments are real and substantial. They're part of why DeepSeek's open-weights releases have been broadly celebrated within parts of the AI research and engineering community.

The Case Against (Or At Least, the Concerns)

The critics of open-weight frontier AI have their own arguments:

Misuse risk. Open-weight models can be modified to remove safety constraints. Capabilities the original developers tried to limit (generating extremist content, aiding bioweapons development, producing CSAM) can be unlocked by determined bad actors. Closed-weight models with API gating are easier to constrain.

Proliferation. Once frontier weights are released, they can't be unreleased. As capability rises, the global floor of what's available rises with it. Concerns about AI capability reaching hostile states or non-state actors shift from "if" to "when."

Race-to-the-bottom dynamics. If open-weight models with weaker safety behavior become widely adopted, commercial pressure on closed-weight providers to match capability may erode safety practices industry-wide.

Accountability gaps. When a closed-weight provider's model causes harm, you know who's responsible — the provider. When an open-weight model causes harm, the chain of responsibility is murky. Who's accountable: the original lab that trained it, the user who deployed it, the platform that hosted the deployment?

Asymmetric global capability. When frontier AI is freely available globally, it's also available to actors with less democratic accountability than typical commercial AI providers. That affects power dynamics in ways that don't have clean answers.

These concerns are also real and substantial. They're part of why open-weight frontier releases have been controversial in AI safety and policy communities.

Where DeepSeek Fits in the Debate

DeepSeek has been at the center of the open-weights debate in concrete ways:

Capability level. DeepSeek's open-weight releases (V3, R1, V4) are at or near the frontier. This isn't "open-source small models for hobbyists" — these compete with the strongest commercial offerings. The capability-proliferation question is sharp.

Safety constraints. DeepSeek's models include some safety training but are less aggressively constrained than some commercial alternatives. Modifications that remove most behavioral guardrails are well-documented in the community.

Geopolitical positioning. A Chinese lab releasing frontier-capability models globally carries strategic implications that purely technical evaluations skip. Both supporters and critics of open weights have specific positions on what this means.

Acceleration. DeepSeek's releases have measurably accelerated the open-weights ecosystem — encouraging other labs to release at similar capability, attracting more talent and resources to open-weight work, normalizing the assumption that frontier capability will be increasingly accessible.

For DeepSeek users, this background isn't abstract — it shapes the broader environment in which DeepSeek and tools like it will continue to develop.

Practical Ethics for DeepSeek Users

Setting aside the broader policy debate, individual users and organizations face concrete ethical questions when using DeepSeek:

Disclosure

When you use AI assistance in producing work, when should you disclose it? The norms are evolving fast. A few principles that hold up:

Disclose for substantial use. AI that produced the content gets disclosed; AI that helped you research or revise may not need explicit disclosure, depending on context.

Disclose for high-stakes use. Customer-facing communications, regulated-industry deliverables, academic publications, anything where the audience would care how the content was produced — disclose.

Match the venue's expectations. Some publications, conferences, and platforms have specific AI disclosure requirements. Comply with what applies to you.

Default to disclosure when uncertain. The downside of disclosing assistance you didn't need to disclose is small. The downside of failing to disclose assistance that turns out to matter can be large.

Verification

We've covered this throughout the book, especially in Chapter 30. The ethical dimension: when you publish work or act on AI output, you're responsible for the accuracy. The model's failure modes don't excuse the human author from due diligence.

For high-stakes work, verification isn't optional — it's the part of the work where you take responsibility.

Attribution

For work that builds substantively on AI output (drafts you edited, code you adapted), attribution practices vary by context:

- **Academic work**: typically a methods-section disclosure, not co-authorship
- **Journalism**: editorial standards have been tightening; consult your publication's policy
- **Software**: open-source projects increasingly disclose AI assistance in commit messages or contribution notes
- **Creative work**: norms are unsettled; honest disclosure of substantive AI involvement is the safe default

The principle: don't claim more originality than you delivered, but don't mistake using a tool for being created by the tool either.

Use of Sensitive Capabilities

Open-weight models can be modified or fine-tuned for purposes the original developers didn't intend. Some are legitimate (specialized domain applications, safety research, education). Others aren't (generating extremist content, automating harassment, producing illegal material).

For ethical users, the principle is straightforward: don't fine-tune or modify models for harmful purposes. The technical capability existing doesn't make using it acceptable.

For organizations deploying open-weight models, the responsibility extends to how the model is deployed, what content it can produce in your application, and what abuse potential the deployment enables.

Provider Choice as Ethical Choice

Choosing which AI provider to use is itself an ethical choice. It affects:

- Who gets paid for AI capability, and what they do with the revenue
- What kind of AI ecosystem gets reinforced (open vs. closed, geographic concentration)
- What political dynamics get supported (covered in Chapter 31)

- What kinds of user data get processed where

For organizations making these choices at scale, the ethical dimensions deserve explicit consideration alongside the technical and economic ones.

The Broader Trajectory

Where the open-weights movement might be heading:

Continued capability releases. The trend toward more frontier-capable open-weight models will likely continue. DeepSeek won't be the last lab to release at this capability level.

Regulatory response. Governments are likely to respond to open-weight AI proliferation with some mix of restrictions, requirements (safety evaluations before release), and investment in their own capabilities.

Safety research adaptation. AI safety research has historically focused largely on closed-weight commercial models. The shift toward open weights is forcing methodology changes — research on how to make capable open-weight models safer, how to constrain their misuse, how to detect modifications.

Industry consolidation around access patterns. A pattern that may emerge: closed-weight commercial models for the highest-stakes regulated workloads, open-weight models for everything else. Or other patterns we can't yet predict.

Geopolitical evolution. The current US-China dynamic around AI is probably not the long-term equilibrium. How it evolves affects everything else.

For users of AI today, the practical implication: the environment will keep changing. Decisions you make now (architectural, contractual, ethical) should be made with awareness that the surrounding context isn't stable.

A Personal Conclusion

This book has tried to give you the working knowledge to use DeepSeek effectively — the technical patterns, the cost optimizations, the operational disciplines, the comparisons. It's also tried to be honest about the harder dimensions: the failure modes, the political context, the ethical questions.

The synthesis: DeepSeek is a remarkable tool that exists within a complicated context. Using it well requires both the practical skills and the broader awareness. The practical skills produce good outputs. The broader awareness keeps you using the tool in ways you can defend openly.

For most users, most of the time, DeepSeek is genuinely useful. Use it for the workloads where it fits. Verify what matters. Disclose where it's expected. Pay attention to the dimensions that aren't purely technical. Adapt as the environment evolves.

That closes Part VIII, and with it the substantive content of the book. The remaining material is reference — the epilogue, the appendices with API quick reference, prompt templates, troubleshooting, glossary, what's-new-in-2026, and resources for further learning.

What you've learned across the preceding eight parts — the mechanics, the workflows, the cost discipline, the ethical posture — is enough to use DeepSeek well, for years, across the kinds of work you actually do.

The rest is practice.

Epilogue: The Cost Curve Broke

The loud AI story of 2024 and 2025 was about capability — each new model release pushed benchmarks, demos went viral, the talk-show coverage focused on what AI could now do that it couldn't before.

DeepSeek's story was different. Quieter at first. V3's capability in December 2024 was competitive but not radically ahead. The breakthrough wasn't what the model could do; it was what it cost to do it. A frontier-class model trained for under $6 million when its Western competitors spent fifty to a hundred million. An API priced at one-tenth what Western providers charged. Open weights that anyone could download and run.

The capability stories will keep coming. Every few months, some lab will release something measurably more impressive. Eventually the pace will slow; eventually new dimensions of impressive will replace the current ones. The capability story is real and important and not going to stop.

But the cost story DeepSeek told in late 2024 and through 2025 may turn out to be more consequential than any single capability release. The economics of AI shifted underneath everyone. Workloads that were too expensive to consider became routine. Architectures that depended on premium API pricing got rethought. The space of economically viable AI-powered things expanded by a factor of ten.

This is the quieter revolution. Less photogenic than Sora videos or AlphaFold protein structures, but more directly consequential for the engineering organizations actually integrating AI into real products.

What DeepSeek Represents Beyond Itself

DeepSeek itself is one company, one product line, one set of strengths and constraints. Even if it evolved or stumbled tomorrow, the patterns it established would persist.

Open-weight frontier models are now the default expectation. Other labs — Meta with Llama, Mistral, Alibaba with Qwen, and others — have been releasing open-weight models at competitive capability. The market is no longer "API-only or nothing" at the frontier.

Cost-per-quality-point dropped permanently. Even if specific DeepSeek pricing rises, the existence proof that capable models can be served at low cost has reset competitive expectations across providers.

Self-hosting is a real option for serious users. Distilled models that run on workstation GPUs put frontier-derived capability in the hands of individuals and small teams without API dependencies.

Reproducibility for AI research became practical. Closed-API-based research was always implicitly irreproducible. Open-weight models change that for the parts of the research community that prioritize reproducibility.

Multi-provider architectures are the new default. Routing between providers based on workload, cost, and capability is now standard practice. Vendor lock-in is a choice you make deliberately, not a default.

None of this is reversible. The shape of the AI industry has shifted. DeepSeek didn't single-handedly cause it all, but it accelerated and crystallized changes that might have happened more slowly otherwise.

What You've Built

If you've read this book and used DeepSeek alongside it, you've built something concrete. Maybe an application that uses DeepSeek for cost-efficient text generation. Maybe a coding workflow that pairs DeepSeek with your IDE. Maybe a self-hosted deployment for privacy-sensitive workloads. Maybe a multi-provider routing layer that pairs DeepSeek with Claude or GPT for the workloads where each is strongest.

Whatever it is, it's yours. Built deliberately, with the cost discipline the API rewards, with the verification habits that protect you from failure modes, with the operational patterns production deployments need.

That's the actual value DeepSeek delivers. Not the headline capability. Not the cost reduction in isolation. The combination of capability, cost, and operational tractability that makes serious AI integration viable for organizations and individuals who couldn't justify it before.

What Comes Next

DeepSeek will keep evolving. New model versions, new specialized models, new pricing structures, new capabilities. The lineup as described in this book is accurate as of early 2026; significant pieces will have changed by the time you read this.

The principles will hold up longer than the specifics. Cost awareness as production discipline. Caching as the high-leverage optimization. Verification as the non-negotiable habit. Multi-provider routing as the resilient architecture. Honest treatment of the political and ethical dimensions. These aren't DeepSeek-specific; they're how to do AI integration well in the open-weights era, regardless of provider.

Beyond DeepSeek, the broader AI landscape will keep shifting. Capabilities will improve. Costs will keep dropping (though probably more slowly than during the V3-to-V4 era). Regulation will catch up to the technology in ways that affect what's possible. The geopolitical dynamics will evolve in directions that are hard to predict.

You're working in the early years of an industry that's still finding its shape. The choices you make about which providers to use, which architectures to build, which ethical positions to hold — these are part of what shapes the industry's eventual form.

A Personal Note

If this book has done its job, you have:

- The technical knowledge to use DeepSeek effectively across the surfaces it offers (web app, API, self-hosted)
- The cost discipline to build production deployments that don't blow up your budget
- The verification habits that protect you from the failure modes that catch careless users
- The architectural awareness to integrate DeepSeek alongside other providers as the workload demands

- The honest framing of the political and ethical dimensions that come with using a Chinese open-weights provider

Most importantly, you have the ability to evaluate, on your own, whether a particular use of DeepSeek is the right use. Not because the book told you the answer for every situation — it didn't, because no book could — but because the principles, frameworks, and habits the book covered let you reason about new situations as they come up.

The features will change. The pricing will shift. The model lineup will evolve. But the way of thinking about the tool — calibrated trust, deliberate cost design, multi-provider awareness, ethical clarity — carries forward.

You're equipped. Now do the work.

Where to Go From Here

If you've found this book useful, the broader Finnoybu Press AI for Everyone series covers complementary topics. *NotebookLM: The Complete Guide* covers Google's source-grounded research tool — the natural complement to DeepSeek for document-heavy workflows. *Cursor: The Complete Guide* covers the AI-native code editor that integrates DeepSeek as one of its supported model providers. *AI Prompt Engineering* covers the deeper craft of prompting across all the major AI tools. *AI Governance for Practitioners* takes the ethical and institutional questions raised in Chapters 31–32 and develops them into a full framework for organizations adopting AI responsibly.

To stay current with DeepSeek specifically, watch the official changelog at api-docs.deepseek.com, the Hugging Face model pages for new releases, and the community on GitHub and Reddit for emerging patterns. Appendix F has more curated resources.

For the broader practice of AI integration and development, you'll do better building your own habits than reading more books on the subject. The skills compound through use, not study. Pick a workflow from Part VI that fits your work, try it for a month, refine based on what you learn.

That's the end of the book. The rest is reference material — API quick sheet, prompt templates, troubleshooting, glossary, what's new, further resources. Use them when you need them. Come back to them as the platform evolves.

Thanks for reading. Now: open the API console, build something useful, and watch your cache hit rates.

Part IX: Reference

"The right reference is the one you can find when you need it."

— Anonymous

Appendix A: API Reference Quick Sheet

A one-stop reference for the DeepSeek API as of early 2026. Specific endpoints and parameters evolve; check api-docs.deepseek.com for current details. The basics below have been stable.

Base URL

```
https://api.deepseek.com/v1
```

Authentication

Bearer token in the Authorization header:

```
Authorization: Bearer sk-...
```

Endpoints

`POST /v1/chat/completions` — primary chat completion endpoint, OpenAI-compatible

`POST /v1/completions` — legacy completions endpoint, used for FIM (fill-in-the-middle)

`POST /v1/embeddings` — embedding generation

`GET /v1/models` — list available models

Chat Completion Request Shape

```
{
  "model": "deepseek-v4-flash",
  "messages": [
    {"role": "system", "content": "..."},
    {"role": "user", "content": "..."}
  ],
  "temperature": 0.7,
```

```
  "max_tokens": 1000,
  "stream": false,
  "response_format": {"type": "json_object"},
  "tools": [...],
  "tool_choice": "auto",
  "extra_body": {"thinking": {"type": "enabled"}}
}
```

Chat Completion Response Shape

```
{
  "id": "chatcmpl-...",
  "object": "chat.completion",
  "created": 1234567890,
  "model": "deepseek-v4-flash",
  "choices": [
    {
      "index": 0,
      "message": {
        "role": "assistant",
        "content": "...",
        "tool_calls": [...]
      },
      "finish_reason": "stop"
    }
  ],
  "usage": {
    "prompt_tokens": 100,
    "completion_tokens": 50,
    "total_tokens": 150,
    "prompt_tokens_details": {
      "cached_tokens": 80
    }
  }
}
```

Common Models (Early 2026)

Model String	Use For
deepseek-v4-pro	Highest capability, complex tasks
deepseek-v4-flash	Cost-efficient default
deepseek-chat	Legacy alias (V3.2); will deprecate

Model String	Use For
deepseek-reasoner	Legacy alias (R1); will deprecate
janus-pro	Multimodal (vision)
deepseek-ocr	Document OCR
deepseek-prover-v2	Formal mathematics (Lean)

Common Parameters

- **`temperature`** — 0.0 (deterministic) to 2.0 (highly random); default 0.7
- **`top_p`** — nucleus sampling cutoff; default 1.0
- **`max_tokens`** — maximum output tokens
- **`stream`** — true for SSE streaming, false for single response
- **`stop`** — array of stop sequences
- **`presence_penalty`** — discourages repeated topics
- **`frequency_penalty`** — discourages repeated tokens
- **`seed`** — for reproducible sampling (best-effort)
- **`response_format`** — `{"type": "json_object"}` for JSON mode, `{"type": "text"}` for free-form
- **`tools`** — array of function definitions for tool calling
- **`tool_choice`** — `"auto"`, `"none"`, or `{"type": "function", "function": {"name": "..."}}`
- **`extra_body.thinking`** — enable thinking mode: `{"type": "enabled"}` or `{"type": "disabled"}`

Tool Call Definition

```
{
  "type": "function",
  "function": {
    "name": "function_name",
    "description": "What the function does",
    "parameters": {
      "type": "object",
      "properties": {
        "arg_name": {"type": "string", "description": "..."}
      },
```

```
      "required": ["arg_name"]
    }
  }
}
```

Tool Call Response

```
"tool_calls": [
  {
    "id": "call_...",
    "type": "function",
    "function": {
      "name": "function_name",
      "arguments": "{\"arg_name\": \"value\"}"
    }
  }
]
```

Tool Result Message

```
{
  "role": "tool",
  "tool_call_id": "call_...",
  "content": "result content as string or JSON"
}
```

FIM Request Shape

```
POST /v1/completions
{
  "model": "deepseek-v4-flash",
  "prompt": "<text before cursor>",
  "suffix": "<text after cursor>",
  "max_tokens": 80,
  "stop": ["\n\n"]
}
```

Note: FIM requires non-thinking mode.

Embedding Request Shape

```
{
  "model": "deepseek-embedding",
```

```
  "input": "text to embed"
}
```

Or for batch embeddings:

```
{
  "model": "deepseek-embedding",
  "input": ["text 1", "text 2", "text 3"]
}
```

Common Error Codes

- **200** — success
- **400** — bad request (malformed input, invalid parameters)
- **401** — unauthorized (bad or missing API key)
- **402** — payment required (out of credit)
- **403** — forbidden (insufficient permissions or restricted region)
- **404** — not found (invalid endpoint or model)
- **422** — unprocessable (validation error in request body)
- **429** — rate limited (back off and retry)
- **500/502/503** — server errors (retry with exponential backoff)

Retry Strategy

For 429 and 5xx errors:

```
import time
import random

def call_with_retry(client, max_retries=5, **kwargs):
    for attempt in range(max_retries):
        try:
            return client.chat.completions.create(**kwargs)
        except (RateLimitError, APIError) as e:
            if attempt == max_retries - 1:
                raise
            time.sleep((2 ** attempt) + random.random())
```

For 4xx errors (other than 429): don't retry; fix the request.

Cost Calculation Reference

Per million tokens, approximate (early 2026):

Model	Input (cache miss)	Input (cache hit)	Output
V4 Pro	$0.43	$0.0036	$0.87
V4 Flash	$0.14	$0.0028	$0.28

Cost per request:

```
def cost_dollars(usage, rates):
    cached = usage.prompt_tokens_details.cached_tokens or 0
    miss = usage.prompt_tokens - cached
    output = usage.completion_tokens
    return (
        cached * rates["cache_hit"] +
        miss * rates["cache_miss"] +
        output * rates["output"]
    ) / 1_000_000
```

Headers Worth Knowing

Request headers: - `Authorization: Bearer sk-...` - `Content-Type: application/json`

Response headers (when present): - `X-RateLimit-Limit-Requests` — your RPM ceiling - `X-RateLimit-Remaining-Requests` — RPM remaining - `X-RateLimit-Limit-Tokens` — your TPM ceiling - `X-RateLimit-Remaining-Tokens` — TPM remaining - `Retry-After` (on 429) — seconds to wait before retrying

URLs to Bookmark

- `api-docs.deepseek.com` — official API documentation
- `platform.deepseek.com` — console for keys, billing, usage
- `chat.deepseek.com` — web chat app
- `huggingface.co/deepseek-ai` — open-weight model releases

For the freshest spec, the official docs always supersede this quick sheet.

Appendix B: Prompt Template Library

Sixty-plus prompt templates organized by use case. Adapt the bracketed variables to your situation. Most are calibrated for the chat-completion API; many work in the web app too with light adaptation.

Coding and Debugging

1. *"Write a [language] function that does [specific task]. Constraints: [list]. Include type hints, docstrings, and minimal example usage."*
2. *"Explain what this code does, line by line: [code]"*
3. *"This code is failing with this error: [error]. Code: [code]. Diagnose the issue and suggest a fix."*
4. *"Refactor this function for [readability / performance / testability]. Preserve the external interface. Code: [code]"*
5. *"Generate [framework, e.g., pytest] tests for this function. Cover happy path, edge cases, and error conditions. Code: [code]"*
6. *"Translate this code from [source language] to [target language]. Preserve behavior; use idioms appropriate to the target language. Code: [code]"*
7. *"Review this PR diff for bugs, style issues, and improvements. Be concrete. Diff: [diff]"*
8. *"Add comprehensive docstrings to these functions. Use [Google / NumPy / Sphinx] style. Code: [code]"*
9. *"This API call is timing out. Code: [code]. What are the most likely causes and what should I check first?"*
10. *"Design a database schema for [use case]. Include table definitions with columns, indexes, and foreign keys. Note the trade-offs you made."*

Architecture and Design

11. *"I'm designing [system]. Constraints: [list]. What's a defensible high-level architecture? What are the major decision points and trade-offs?"*

12. *"Compare these three approaches to [problem]: [option A], [option B], [option C]. Identify the situations where each wins."*

13. *"Walk me through how I'd build [system] using [tech stack]. Identify the major components and how they connect."*

14. *"What's the right error-handling strategy for [system]? Be specific about what to retry, what to fail fast, and what to escalate."*

15. *"Design the API surface for [feature]. Show endpoint paths, methods, request/response shapes, and the most important error cases."*

Reasoning and Analysis

16. *"Walk through your reasoning step by step before giving the final answer. Question: [question]"*

17. *"Steel-man both sides of this debate: [topic]. What's the strongest case for each position?"*

18. *"What are the most likely failure modes of this approach: [approach]? Rank by likelihood and impact."*

19. *"Identify the implicit assumptions in this argument: [argument]. Which assumptions, if false, would invalidate the conclusion?"*

20. *"What's the strongest counter-argument to this position: [position]? Be specific."*

Math and Quantitative

21. *"[problem]. Walk through the setup, the math, and the answer. Show your work."*

22. *"Verify this calculation: [calculation]. Identify any errors. If correct, confirm with explicit reasoning."*

23. *"Translate this word problem into a formal mathematical statement: [problem]"*

24. *"What probability distribution is appropriate for [scenario]? Justify the choice."*

25. *"Estimate [quantity] using only the following inputs: [inputs]. Show the calculation."*

Research and Synthesis

26. *"Summarize the key findings from this document in three paragraphs: [document]"*

27. *"Compare the methodologies used in these papers: [papers]. Identify strengths and weaknesses of each."*

28. *"What questions does this paper leave unanswered: [paper]?"*

29. *"Identify the strongest piece of evidence for this claim and the strongest evidence against it: [claim]"*

30. *"Find the central thesis of this document and assess how well the body of the document supports it. Document: [document]"*

Writing and Editing

31. *"Rewrite this in [target voice / register / tone]. Preserve the meaning. Original: [text]"*

32. *"Edit this for clarity. Don't change my voice. Original: [text]"*

33. *"Cut this by [N]%. Preserve the most important content. Original: [text]"*

34. *"Suggest five different headlines for this article: [article]. Each should target a different audience or angle."*

35. *"Translate this from [source language] to [target language]. Preserve tone and intent, not just literal meaning. Original: [text]"*

Structured Extraction (use with JSON mode)

36. *"Extract the following fields from the input as JSON: [field list with descriptions]. Use null for missing fields. Input: [text]"*

37. *"Classify this input into one of these categories: [list]. Respond as JSON:* `{\"category\": \"...\", \"confidence\": 0.0-1.0, \"rationale\": \"...\"}`*. Input: [text]"*

38. *"Extract every [entity type] mentioned in the text. Respond as JSON array of objects with [field list]. Text: [text]"*

39. *"Identify the sentiment of this text on each of these dimensions: [dimensions]. Respond as JSON. Text: [text]"*

40. *"Parse this unstructured input into the schema: [schema]. Respond as JSON conforming to the schema. Input: [input]"*

Tool Use (Agent Patterns)

41. *"Use the available tools to: [objective]. Plan your steps, then execute. If a tool fails, decide whether to retry or try a different approach."*

42. *"Investigate [topic] using the available tools. Stop when you have enough information to write a one-page report."*

43. *"Find [specific information] using web search. Cite sources. If the information isn't available, say so explicitly."*

44. *"Walk through this customer's account history using the database tool. Identify any anomalies."*

45. *"Read the relevant files using the file system tools, then propose a fix for [issue]."*

Cost-Optimized (Cache-Friendly Patterns)

46. *(System prompt, stable):* "You are an expert [role]. Your output should follow this structure: [structure]. Constraints: [constraints]."

47. *(System prompt, stable):* "Process the user input according to the rules below. [Rules]. Respond as JSON: [schema]."

48. *(User prompt template):* "Input to process: [the only variable part]"

49. *(For RAG):* `[STABLE: system prompt]` + `[VARIABLE: retrieved context]` + `[VARIABLE: user question]` — context first, question last to maximize prefix caching.

50. *(For chat with history):* The conversation history naturally creates a growing cacheable prefix; just keep appending rather than reordering.

Verification and Quality Checks

51. *"Review the previous answer for accuracy. Identify any specific claims that should be verified, and explain how to verify each."*

52. *"Where in your previous answer were you uncertain? Be honest about confidence levels."*

53. *"Imagine this answer is wrong. What's the most likely error?"*

54. *"This is the answer: [answer]. Critique it. What would a domain expert push back on?"*

55. *"Verify this code by walking through it with these specific test inputs: [inputs]. What output would each produce?"*

Self-Correction Patterns

56. *"You said [claim]. Source [document] suggests this is incorrect. Re-examine and revise if needed."*

57. *"That's not quite right. The actual situation is: [correction]. Now answer the original question accurately."*

58. *"Try a different approach. The previous one [didn't work / was inefficient / produced wrong results]."*

59. *"Reconsider with this additional constraint: [constraint]. Update your answer accordingly."*

60. *"Double-check the arithmetic in your previous answer. If correct, confirm. If not, fix."*

Reflection and Synthesis

61. *"Summarize what we've established in this conversation. What are the key conclusions and what's still uncertain?"*

62. *"Looking back at this discussion, what's the most important point to remember?"*

63. *"What questions should I ask that I haven't asked yet, given what we've discussed?"*

64. *"Identify the three most consequential decisions we've made in this conversation. For each, note the reasoning and the alternatives we considered."*

65. *"Take everything we've covered and condense it into a single paragraph someone could read in 30 seconds."*

General-Purpose Power Patterns

For most use cases, these patterns punch above their weight:

- **Be specific about format.** "Three paragraphs. Two sentences each. Plain text, no headers." beats "Write a short response."
- **Specify the audience.** "Explain to a non-technical executive" produces different output from "explain to a senior engineer."
- **Give explicit constraints.** "Don't use jargon. Don't include caveats. Don't suggest alternatives I didn't ask about."
- **Ask for the work, not just the answer.** Especially for high-stakes calls, requesting reasoning gives you something to verify.
- **Iterate with corrections rather than restarting.** "Now do the same but for [different audience]" beats restarting from scratch.

Appendix C: Troubleshooting Common Issues

Quick-reference solutions for the issues DeepSeek users encounter most often. Organized by symptom.

Authentication and Access

`401 Unauthorized`. - Verify your API key is correct (no extra whitespace, complete value) - Confirm you're using `Authorization: Bearer sk-...` (Bearer prefix is required) - Check the key hasn't been revoked (console → API Keys) - Verify you're using the correct endpoint URL (`https://api.deepseek.com/v1`)

`402 Payment Required`. - Your account is out of credit. Top up via the console. - For paid accounts on auto-recharge, verify the payment method is current

`403 Forbidden`. - Geographic restriction may apply (some regions are unavailable) - Account may have been restricted; contact DeepSeek support

Account locked or suspicious activity. - Often triggered by IP changes, bursts of failed authentication, or unusual usage patterns - Reach out to DeepSeek support; provide your account email and recent activity context

Rate Limits

`429 Too Many Requests`. - You've exceeded RPM (requests per minute) or TPM (tokens per minute) limits - Implement exponential backoff (retry with increasing delays: 1s, 2s, 4s, 8s, ...) - Consider request batching to reduce per-minute call count - For sustained heavy load, request higher rate limits via DeepSeek's commercial team

Limits hit despite low overall traffic. - Burst traffic can hit RPM even when average traffic is low - Add a request queue with deliberate pacing - Check for retry storms (aggressive retries on failed requests can multiply traffic)

Model Behavior Issues

Model returns refusal on a topic you expected it to handle. - Politically sensitive topics (Tiananmen, Xinjiang, Taiwan, etc.) trigger consistent refusals on hosted services (Chapter 31) - Try rephrasing the question; sometimes the refusal is triggered by specific keywords rather than the topic itself - For unrestricted use, consider self-hosting the open weights

Output is shorter than requested. - Verify `max_tokens` is set high enough for the requested length - Check `finish_reason` in the response — `length` means the model hit the cap - For long outputs, request the content in sections rather than all at once

Output is much longer than requested. - Be specific about length in the prompt: "in 200 words", "as three sentences", "no more than 500 tokens" - Set `max_tokens` to enforce a hard ceiling - Add `stop` sequences for structured outputs (e.g., `stop=["\n\n"]` to terminate at paragraph breaks)

Output quality lower than expected. - Try thinking mode for the call (`extra_body={"thinking": {"type": "enabled"}}`) - Try a more capable model (V4 Pro instead of V4 Flash) - Improve the prompt: more specific instructions, examples, format constraints - Check the system prompt for conflicts or unclear directives

JSON mode returns invalid JSON. - Shouldn't happen with `response_format: {"type": "json_object"}` set, but if it does: - Verify the prompt explicitly asks for JSON output (the system shouldn't have to infer it) - Check that the prompt provides the schema or output structure - Wrap the parse in a try/except and retry once

Same prompt returns different outputs each time. - Default temperature is non-zero; outputs are stochastic by design - Set `temperature: 0` for more deterministic output (still not perfectly reproducible) - Set `seed` parameter for best-effort reproducibility

Tool Calling Issues

Model doesn't call any tools when you expected it to. - Tool descriptions may be unclear; the model can't tell when each applies - The system prompt may have discouraged tool use; verify it doesn't say "answer directly without using tools" - Try thinking mode — explicit reasoning improves tool selection

Model calls the wrong tool. - Tool descriptions need to clearly differentiate when each applies - Reduce the tool count if too many overlapping options confuse the model - Add explicit guidance in the system prompt about when to use which tool

Model returns invalid JSON in tool arguments. - JSON Schema enforcement helps but isn't always strict - Validate tool arguments in your execution code with try/except - For critical parameters, fail loudly — don't proceed with default values silently

Tool execution failures crash the agent. - Wrap tool calls in try/except - Return errors to the model as structured tool results (`{"error": "..."}`) so it can react - Set hard step limits to prevent runaway loops on persistent failures

Streaming Issues

Stream gets stuck or hangs. - Network interruption — check connectivity - Verify the OpenAI client is set up correctly for streaming (`stream=True`) - Set read timeouts on the stream so a stuck connection doesn't hang forever

Streaming chunks have empty content. - Normal — some chunks (especially for tool calls) have no text content but other delta fields - Check `chunk.choices[0].delta.content` is truthy before appending

Token counts unavailable in streaming responses. - Token usage is typically reported only in the final chunk (or in a separate field) - Some streaming implementations omit usage entirely; you may need to estimate or count locally

Cost and Caching Issues

Cache hit rate lower than expected. - Check the order of content in your prompts: stable content must be at the start (cache prefix matches from the beginning) - Look for variable content sneaking into the prefix (timestamps, request IDs, dates) - Verify the same prompt structure across calls — small variations bust the cache

Costs spiked unexpectedly. - Check whether thinking mode got enabled where it wasn't expected - Look for runaway agent loops (long sessions with many tool calls) - Verify model selection — accidentally using V4 Pro instead of V4 Flash multiplies cost - Check for prompt drift — gradually more verbose prompts increase token volume

Token counts in usage don't match expectations. - Tokenization isn't 1:1 with words; expect roughly 0.75 word per token for English text - Code and structured data tokenize less efficiently - Use `tiktoken` or similar locally to verify expected token counts

Self-Hosting Issues

Out of memory loading model. - Model is too large for your GPU. Pick a smaller variant or use more aggressive quantization - Verify you're not loading multiple models simultaneously - Consider tensor parallelism across multiple GPUs

Inference very slow. - May be running on CPU when you expected GPU. Check `nvidia-smi` (Linux/Windows) or Activity Monitor (Mac) - Update GPU drivers and CUDA toolkit - For Ollama specifically, verify the GPU is being detected (`ollama show <model>` shows hardware info)

Output differs between local and API. - Quantization affects quality; FP16 matches the API more closely than Q4 - Confirm you're running the version you expect locally - Sampling parameters (temperature, top_p) may default differently between local and API

vLLM crashes on startup. - Verify CUDA and driver versions match vLLM's requirements - Check available memory before launch — vLLM is memory-greedy by default - Reduce `--gpu-memory-utilization` if other processes share the GPU

FIM completion produces wrong output. - Verify you're using the completions endpoint (`/v1/completions`), not chat completions - Check that thinking mode is OFF (FIM requires non-thinking) - Make sure prefix and suffix are in the correct fields (`prompt` for prefix, `suffix` for suffix)

Operational Issues

Intermittent timeouts. - DeepSeek's API has variable latency; set your client's request timeout generously (60–120s for non-thinking, 300+ for thinking-mode calls on long outputs) - Add retry logic with backoff for transient failures - For latency-critical applications, consider falling back to a faster model on slow responses

Webhook events not arriving. - DeepSeek doesn't have webhook events for chat completions; if you're seeing this, you may be confusing it with another service - For Stripe-style webhooks (e.g., billing events), verify the webhook URL is configured and your endpoint is reachable

Embedded image not being read. - Verify the image is base64-encoded correctly and the data URL prefix is right (`data:image/png;base64,` or appropriate MIME type) - Check image size; very large images may be rejected - Some image formats are better supported than others (PNG and JPG are the most reliable)

When to Contact Support

Most issues resolve through the troubleshooting above. For issues that persist:

- **Account-specific problems** — billing, key issues, region availability
- **Suspected model bugs** — reproducible failures that look like model errors rather than user errors
- **Persistent infrastructure issues** — sustained latency or error rates beyond what's documented as normal
- **Enterprise needs** — higher rate limits, custom contracts, dedicated support

DeepSeek's support channels are documented at api-docs.deepseek.com. For most issues, the community on GitHub and Reddit is also a useful first stop — others have likely hit the same issue.

For most everyday problems, the fix is one of: refresh credentials, retry with backoff, rephrase the prompt, switch models, verify caching, or check the docs for changes since you last looked.

Appendix D: Glossary of Terms

A reference for the key terms used throughout this book and across DeepSeek's documentation, ecosystem, and broader open-weights AI discourse.

Agent. A language model in a loop, taking actions via tools and adapting based on results. Distinguished from chains (fixed sequences of calls) by the model's autonomy in deciding what to do next.

Agent loop. The pattern of repeatedly calling the model, executing tools it requests, feeding results back, and continuing until the model decides the task is complete or a limit triggers.

API key. A string credential identifying your account for authenticated API requests. Stored in environment variables or secrets management; never in source code.

Attention. The neural-network mechanism that lets the model weight different parts of its input differently when generating each output token. Multiple attention variants exist (standard, sparse, compressed, etc.).

BentoML. A production-grade serving framework for ML models, including LLMs. Wraps inference engines like vLLM with deployment, monitoring, and scaling.

Cache hit. An input token that matches a previously-cached prefix from a recent request, charged at a deeply discounted rate (typically 80–90% off).

Cache miss. An input token that doesn't match a cached prefix, charged at the standard input rate.

Chat completion. The primary API endpoint shape for conversational AI: a list of messages in/out. The OpenAI-defined shape that most providers (including DeepSeek) implement.

chat.deepseek.com. DeepSeek's web chat application, free for individual use.

Compressed Sparse Attention. Attention mechanism used in DeepSeek-V4 for efficient long-context handling. Reduces compute and memory required for the 1M-token context window.

Context window. The amount of text (input + output) the model can consider at once. DeepSeek-V4 has a 1M-token context window.

Distillation. Training a smaller "student" model to mimic the behavior of a larger "teacher" model. DeepSeek's distilled lineup uses R1 as the teacher.

Embedding. A vector representation of text used for similarity search, retrieval, and other ML pipelines. Different from chat completions; produced by dedicated embedding endpoints.

FIM (Fill-In-the-Middle). A completion mode where the model fills in content between a given prefix and suffix. Used for code autocomplete and document infilling.

Hallucination. Model output that's plausible but factually wrong — invented facts, fictitious citations, made-up API methods, etc.

Janus. DeepSeek's multimodal vision model lineup. Handles image inputs and (some variants) image outputs.

JSON mode. A response format setting (`response_format: {"type": "json_object"}`) that constrains model output to valid JSON.

KV cache. The key-value cache that holds the model's processed context for the current generation. Memory-managed by inference engines (vLLM's PagedAttention is one such manager).

Lean. A formal proof assistant. DeepSeek-Prover-V2 generates Lean-syntax proofs.

LoRA (Low-Rank Adaptation). A parameter-efficient fine-tuning technique. Adds small trainable adapters to a base model rather than retraining all weights.

Mathlib. The community library of formalized mathematics in Lean. Prover-V2 was trained heavily on mathlib.

Mixture-of-Experts (MoE). An architecture where the model has many "expert" sub-networks but activates only a few per token. DeepSeek's V3 and V4 use MoE designs (V4 Pro: 1.6T total/49B active per token).

Multi-token Prediction (MTP). A training-time technique where the model predicts multiple future tokens jointly, used in some DeepSeek model architectures.

Ollama. Open-source tool for running language models locally, with a simple CLI and HTTP API. The standard entry point for self-hosted DeepSeek.

OpenAI-compatible. APIs that accept requests in the shape OpenAI's API uses. DeepSeek's API is OpenAI-compatible, meaning OpenAI client libraries work against it with one config change.

Open weights. Model parameters released publicly, typically under permissive license. Distinguishes from closed-weight models that are accessible only through provider APIs.

PagedAttention. vLLM's technique for managing the KV cache in chunks (pages), enabling much higher concurrent request batching than naive memory management.

Prompt caching. The mechanism by which the API server caches processed state for prompt prefixes, billing matched tokens at a discounted rate. Most consequential cost optimization for production DeepSeek use.

Prover-V2. DeepSeek's specialized model for formal mathematical proof generation in Lean.

Quantization. Reducing the numerical precision of model weights (FP16 → FP8 → INT8 → INT4) to save memory at some quality cost. FP8 on Hopper GPUs is the typical production sweet spot.

R1 (DeepSeek-R1). DeepSeek's reasoning-focused model, released January 2025. Now largely superseded by V4's thinking mode but still relevant for the distilled-model lineage trained on R1's outputs.

RAG (Retrieval-Augmented Generation). Pattern where relevant documents are retrieved and passed to the model as context, grounding responses in specific source material.

Rate limit. A cap on how many requests or tokens you can submit per unit time. DeepSeek has both RPM (requests per minute) and TPM (tokens per minute) limits.

Reasoning trace. The chain-of-thought output produced by reasoning models (V4 with thinking mode, R1) before the final answer. Visible in a separate field of the response.

Routing. The pattern of deciding which model (or provider) handles each request based on its characteristics. Heuristic, classifier-based, cascade, and tier-based routing are common patterns.

Self-hosting. Running models on your own infrastructure rather than via a hosted API. Enabled by DeepSeek's open-weight releases.

Streaming. Receiving model output as it's generated, token by token, rather than waiting for the complete response. Improves perceived latency for interactive applications.

System prompt. The first message in a chat conversation, typically containing instructions, persona, and behavior constraints. Highly cacheable when stable across requests.

Tensor parallelism. Sharding a model across multiple GPUs by splitting each layer's tensors. Used for serving models too large for a single GPU. Lower latency than pipeline parallelism; requires high-bandwidth GPU interconnect.

Thinking mode. An optional mode (per-call or per-message) where DeepSeek-V4 produces explicit chain-of-thought reasoning before its final answer. Costs more output tokens; improves quality on reasoning-bound tasks.

Token. The unit of text the model processes. Roughly 0.75 word per token in English. Both input and output are billed per token.

Tool calling. The protocol where the model returns a structured request to call a function rather than producing a normal text response. Your code executes the function and returns the result for the model to use.

V3 (DeepSeek-V3). DeepSeek's December 2024 frontier release that triggered the cost-curve recognition. 671B parameter MoE; trained for ~$5.6M.

V4 (DeepSeek-V4). DeepSeek's April 2026 flagship line. Two variants: V4 Pro (1.6T total/49B active) and V4 Flash (284B total/13B active). Built-in thinking mode.

Vision. Multimodal capability handling images as input (and sometimes output). DeepSeek's vision capabilities live primarily in the Janus lineage.

vLLM. Open-source high-throughput inference engine for LLMs. The standard production inference engine for self-hosted deployments. Key innovation: PagedAttention.

Weights. The trained numerical parameters of the model. "Open weights" means publicly downloadable and runnable on your own infrastructure.

Appendix E: What's New in 2026

A timeline of the major DeepSeek developments leading up to and through early 2026, focused on what's changed since V3 released in December 2024.

The Pre-2026 Foundation

2023 — DeepSeek (Hangzhou DeepSeek Artificial Intelligence Co.) founded by Liang Wenfeng, spun off from quantitative hedge fund High-Flyer.

Late 2023 — DeepSeek-V1 released. Competent but not industry-shifting.

Mid-2024 — DeepSeek-V2 released. Continued capability development; modest industry attention.

December 2024 — DeepSeek-V3 released. 671B-parameter MoE architecture with 37B active per token. Trained for approximately $5.6M (vs. ~$50–100M estimates for GPT-4-class Western models). Open-weight release under MIT license. The cost-curve recognition begins.

2025: The Year of Recognition

January 20, 2025 — DeepSeek-R1 released. Reasoning-specialized model competitive with OpenAI's o1 series. Open-weight, MIT-licensed, trained for ~$294K.

January 27, 2025 — Stock-market response. Nvidia drops nearly $600B in market cap on a single day (the largest single-day market-cap loss for any company on record). Mainstream-press AI coverage shifts from "what can AI do" to "what does AI cost."

January–February 2025 — Distilled R1 models released (1.5B through 70B parameters). Self-hosted DeepSeek capability becomes accessible to individual developers.

February–March 2025 — Industry reaction: multiple Western AI providers respond with pricing adjustments, cost-focused product announcements, and (in some cases) hints at open-weight strategies.

March 2025 — DeepSeek-V3-0324 refinement released. Quality improvements on the V3 architecture.

April 2025 — DeepSeek-Prover-V2 released. 671B-parameter MoE model (37B active, 256 experts) specialized in formal theorem proving in Lean 4. Achieves 88.9% on MiniF2F-test.

May 28, 2025 — DeepSeek-V3.1 released, alongside the **DeepSeek-R1-0528** update.

August 21, 2025 — DeepSeek-V3.1-Terminus released. Faster, better tool use, refinements in agent-friendly behaviors. (V3.1-Terminus remained on a temporary API endpoint until October 15, 2025.)

September 29, 2025 — DeepSeek-V3.2-Exp released. Introduces **DeepSeek Sparse Attention (DSA)** for more efficient long-context handling.

December 1, 2025 — DeepSeek-V3.2 launched as the official successor to V3.2-Exp on App, Web, and API. **DeepSeek-V3.2-Speciale** released the same day as a reasoning-focused variant (API-only, available through December 15, 2025).

Throughout 2025 — Janus multimodal vision lineup matures (Janus-Pro-7B becomes the flagship open-weight multimodal release).

2025 — Ecosystem development: - vLLM and BentoML integrations mature for production deployment - Ollama adds DeepSeek models to its registry - Major IDE-AI tools (Cursor, Continue.dev, Aider) add DeepSeek as a supported provider - Multi-provider routing patterns become standard architecture for production AI systems

2026: V4 and Maturity

January 27, 2026 — DeepSeek-OCR-2 released. 3B-parameter vision-language model with the new **DeepEncoder V2** architecture introducing a "causal flow" approach to visual encoding. Achieves 91.09% on OmniDocBench v1.5 (a 3.73% improvement over the original DeepSeek-OCR).

April 24, 2026 — DeepSeek-V4 released as preview. Two variants: **V4 Pro** (1.6T total parameters, 49B active per token) and **V4 Flash** (284B total, 13B active). Both ship with: - 1M-token context window - 384K-token max output -

Built-in optional thinking mode (subsumes R1's reasoning capability) - **Compressed Sparse Attention (CSA)** and **Heavily Compressed Attention (HCA)** hybrid for long-context efficiency - MIT license for open weights - Tool calling, JSON mode, FIM completion (FIM requires non-thinking mode)

At the 1M-token context setting, V4 Pro requires only ~27% of the single-token inference FLOPs and ~10% of the KV cache compared with V3.2.

April–May 2026 — Promotional pricing on V4 Pro: 75% discount through May 31, 2026. Effective rates land V4 Pro at roughly 1/7 the cost of Claude Opus 4.7 on coding workloads.

April 2026 — Thinking mode API surface stabilizes around `extra_body={"thinking": {"type": "enabled"}}` on the OpenAI-compatible endpoint, with `reasoning_content` as the response field. (Multi-turn gotcha: strip `reasoning_content` from prior assistant messages on subsequent input — except after a tool call, where it must be retained.)

Early 2026 — Legacy aliases (`deepseek-chat`, `deepseek-reasoner`) marked for eventual deprecation. Encouragement to migrate to explicit version names (`deepseek-v4-pro`, `deepseek-v4-flash`).

Through early 2026 — Continued refinements: - Janus multimodal updates - Distilled-from-V4 models begin appearing - API stability improvements - Documentation expansion

Trends That Will Likely Continue

A few patterns to watch as the platform continues to evolve:

Continued cost reductions. Each new release has tended to push prices down or hold them while improving capability. Don't assume current prices are floors.

Architectural efficiency improvements. Each new attention mechanism (sparse, compressed sparse, heavily compressed) extends what's tractable at the long-context end.

Reasoning becoming default. Thinking mode is moving from optional to often-default. Expect future releases to make reasoning cheaper and more transparent.

Specialization expansion. Beyond Prover-V2, Janus, and OCR, expect more domain-specific variants (possibly for code, science, security).

Open-weight ecosystem maturation. Distilled variants, fine-tuned community releases, integration tooling — the ecosystem around DeepSeek's weights keeps developing.

Geopolitical and regulatory adaptation. US-China AI dynamics continue to evolve. The regulatory environment around open-weight frontier models is fluid; tracking is worthwhile.

What to Watch For

For users wanting to track DeepSeek's continued evolution:

- **api-docs.deepseek.com** — official API documentation; first place new features land
- **huggingface.co/deepseek-ai** — model releases (the moment new weights are public)
- **DeepSeek's official blog and announcements** — strategic releases, capability claims
- **Independent analysis** in newsletters and AI-research blogs as new releases land
- **Community discussion** on GitHub (vLLM, llama.cpp, Ollama for inference-framework integration), Reddit, and AI-focused forums

The product as described in this book is accurate as of early 2026. Significant changes will appear before the next edition. Use the principles and frameworks in the book to evaluate new features as they arrive — most of what makes DeepSeek useful (or problematic) doesn't change with new releases.

Appendix F: Resources and Further Reading

A curated list of resources for staying current with DeepSeek specifically and the open-weights AI ecosystem more broadly. URLs and specific recommendations were accurate as of early 2026 but evolve quickly; treat this as a starting point.

Official DeepSeek Resources

API Documentation — `api-docs.deepseek.com`. The authoritative reference for current API features, parameters, and pricing. First place new capabilities land.

Console — `platform.deepseek.com`. API key management, billing, usage analytics.

Web App — `chat.deepseek.com`. The free web interface; useful for both casual use and quick capability testing.

Hugging Face Page — `huggingface.co/deepseek-ai`. Model releases, model cards, and community discussion. The moment new open weights are available.

GitHub — DeepSeek's official repositories on GitHub. Source for some of the supporting tooling and reference implementations.

Official Blog and Announcements — DeepSeek's blog and social channels. Strategic announcements and capability claims direct from the source.

Inference and Deployment Tooling

Ollama — `ollama.com`. The standard tool for local LLM inference. Includes pre-packaged DeepSeek model variants.

vLLM — `github.com/vllm-project/vllm`. Production-grade inference engine for high-throughput serving of large language models. Supports DeepSeek's full lineup.

BentoML — `bentoml.com`. Production serving and deployment framework that wraps inference engines like vLLM. Useful for going from "model running on a server" to "model deployed reliably in production."

Hugging Face Transformers — `github.com/huggingface/transformers`. The foundational library for loading and running open-weight models in Python. Necessary for fine-tuning, custom inference, and research workflows.

llama.cpp — `github.com/ggerganov/llama.cpp`. C++ inference engine optimized for CPU and modest hardware. Used by Ollama and others as the underlying engine for many local-inference workloads.

TGI (Text Generation Inference) — `github.com/huggingface/text-generation-inference`. Hugging Face's production inference server, an alternative to vLLM for some workloads.

Multi-Provider and Routing

LiteLLM — `litellm.ai`. Provider abstraction library that lets you call DeepSeek, OpenAI, Anthropic, and others through a uniform interface. Useful for multi-provider routing.

OpenRouter — `openrouter.ai`. A unified API gateway that aggregates many AI providers (DeepSeek included) under a single endpoint. Simpler than building your own abstraction layer for many use cases.

IDE and CLI Integrations

Cursor — `cursor.com`. AI-native code editor; supports DeepSeek as a configured provider via custom OpenAI-compatible endpoint. Covered in detail in *Cursor: The Complete Guide*.

Continue.dev — `continue.dev`. Open-source code assistant for VS Code and JetBrains. First-class DeepSeek support.

Aider — `aider.chat`. CLI-based AI pair programming. Supports DeepSeek as a model provider.

OpenCode — Newer CLI coding agent with first-class DeepSeek support.

Fine-Tuning Frameworks

axolotl — `github.com/OpenAccess-AI-Collective/axolotl`. High-level fine-tuning framework for open-weight models including DeepSeek's distilled lineup.

unsloth — `github.com/unslothai/unsloth`. Memory- and speed-optimized fine-tuning for consumer GPUs. Particularly useful for fine-tuning the smaller distilled variants.

peft (Hugging Face PEFT) — `github.com/huggingface/peft`. Library for parameter-efficient fine-tuning techniques (LoRA, QLoRA, etc.).

Tokenization and Cost Estimation

tiktoken — `github.com/openai/tiktoken`. OpenAI's tokenizer library. Works for DeepSeek's models too (compatible tokenization). Useful for local token counting before sending requests.

Hugging Face Tokenizers — Native tokenizer libraries for non-OpenAI-compatible models, including some DeepSeek variants.

People and Communities

Steven Johnson — Author and AI commentator; thoughtful writing on the open-weights movement and AI's broader trajectory.

Independent AI analysts on Substack, Twitter/X, and Mastodon — a rotating set of commentators who track AI tool development closely. Names change; identify a few whose perspective you trust and follow them.

r/LocalLLaMA — Reddit community focused on running language models locally. Active discussion of DeepSeek deployment, fine-tuning, and use patterns.

r/MachineLearning — Broader ML research community; DeepSeek releases get serious technical discussion here.

Hacker News — General tech news with substantial AI coverage; major DeepSeek releases trigger active comment threads.

Hugging Face community discussions — Model-specific discussion attached to each released model on Hugging Face. Often the highest-quality place for specific technical questions about a model's behavior.

Newsletters Worth Following

The newsletter landscape changes quickly, but in early 2026 the strongest options include:

- **AI tool-tracking newsletters** that cover DeepSeek alongside other providers
- **AI engineering newsletters** focused on production patterns
- **Open-source AI newsletters** that track the open-weights ecosystem specifically
- **Cost-and-capability analysis newsletters** for engineering leaders evaluating providers

Pick two or three that match your specific interests; subscribing to too many produces noise without insight.

Recommended Reading on Open-Weight AI Generally

For readers who want to understand the broader context of what DeepSeek represents:

- **Foundational papers on Mixture-of-Experts** architectures, including DeepSeek's own papers on V3 and V4. Searchable on arXiv.
- **Papers on attention efficiency** (DeepSeek Sparse Attention, Compressed Sparse Attention, FlashAttention, PagedAttention).
- **AI safety research on open-weight models** — papers and posts examining the misuse risk and safety implications of releasing frontier weights publicly.
- **Policy writing on AI export controls and open-weights regulation** — for context on the geopolitical dimensions covered in Chapter 31.

Other Books in the Finnoybu Press AI for Everyone Series

For readers who found this book useful, the broader series covers complementary topics:

- *ChatGPT: The Definitive Guide* — General-purpose AI chatbot. The Western incumbent.
- *ChatGPT: The Power User Guide* — Advanced techniques for ChatGPT specifically.
- *Claude: The Definitive Guide* — Anthropic's general-purpose AI. Strong on long documents and nuanced reasoning.
- *Claude Code: The Power User Guide* — Claude's agentic coding tool.
- *Gemini: The Complete Guide* — Google's general-purpose AI.
- *Gemini: The Advanced Guide* — Power-user techniques for Gemini.
- *Copilot: The Workplace AI Guide* and *Copilot: The Advanced Workplace Guide* — Microsoft's enterprise AI assistant.
- *Perplexity: The Research Playbook* and *Perplexity: The Advanced Research Playbook* — Perplexity's search-and-research focused AI.
- *NotebookLM: From Sources to Audio Overviews* — Google's source-grounded research tool.
- *Cursor: The Complete Guide* — AI-native code editor.
- *OpenClaw: The Complete Guide* — Personal AI assistant.
- *AI Systems Playbook* — Cross-platform patterns for working with multiple AI tools.
- *AI Prompt Engineering* — The deeper craft of prompting across all major AI tools.
- *AI for Developers* — AI tool use for software developers specifically.
- *AI for Business* — AI tool adoption for business teams.
- *AI for Students* — AI tool use across academic contexts.
- *AI Governance for Practitioners* — Frameworks for organizations adopting AI responsibly.
- *Securing Agentic AI* — Security considerations for autonomous AI systems.

Where to Start Each Resource Type

For day-to-day API updates, the official documentation and changelog are sufficient. For deeper context, follow a couple of independent voices and one or two newsletters. For technical depth on the model architecture, read the official DeepSeek papers and the broader literature on MoE architectures and efficient attention.

The most important resource, ultimately, is your own evolving practice. The DeepSeek you use in two years won't be the DeepSeek described in this book, and the workflows that work best for you will be the ones you've refined through your own use. Read, follow, listen — and then build, verify, and refine.

www.ingramcontent.com/pod-product-compliance
Lightning Source LLC
LaVergne TN
LVHW060026210926
848481LV00013B/3794
9798950533112